Louis Bachelier's *Theory of Speculation*

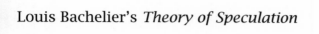

Louis Bachelier's
Theory of Speculation

THE ORIGINS OF MODERN FINANCE

Translated and with Commentary by
Mark Davis and *Alison Etheridge*

PRINCETON UNIVERSITY PRESS
PRINCETON AND OXFORD

ISBN-13: 978-0-691-11752-2 (alk. paper)
ISBN-10: 0-691-11752-7 (alk. paper)

Library of Congress Control Number: 2006928917

A catalogue record for this book is available from the British Library

This book has been composed in Lucida

Typeset by T&T Productions Ltd, London

Printed on acid-free paper ∞

www.pupress.princeton.edu

Printed in the United States of America

10 9 8 7 6 5 4 3 2 1

Contents

Foreword

Mathematical and other scientific research can sometimes have a beauty akin to artistic masterworks. And, rarely, romance can even arise in how science progresses. One notable example of this is the 1914 discovery by the eminent British mathematician, G. H. Hardy, of the unknown mathematical genius Ramanujan—what Hardy called the greatest romance in his professional life.

That story began when the morning post brought to Trinity College, Cambridge, a letter from an unknown impoverished Madras clerk. Instead of giving it a cursory glance, Hardy examined with amazement the several enclosed infinite series. In a flash, he realized they were the work of a genius: a few Hardy already knew; maybe at least one was imperfect; but several were so novel that no mere talent could have discovered them. No need here to rehash the story of the Ramanujan–Hardy collaboration and the tragic early death from tuberculosis of Ramanujan.

As told in the preface below, discovery or rediscovery of Louis Bachelier's 1900 Sorbonne thesis, 'Théorie de la spéculation', began only in the middle of the twentieth century, and initially involved a dozen or so postcards sent out from Yale by the late Jimmie Savage, a pioneer in bringing back into fashion statistical use of Bayesian probabilities. In paraphrase, the postcard's message said, approximately, 'Do any of you economist guys know about a 1914 French book on the theory of speculation by some French professor named Bachelier?'

Apparently I was the only fish to respond to Savage's cast. The good MIT mathematical library did not possess Savage's 1914 reference. But it did have something better, namely Bachelier's original thesis itself.

I rapidly spread the news of the Bachelier gem among early finance theorists. And when our MIT PhD Paul Cootner edited his collection of worthy finance papers, on my suggestion he included an English version of Bachelier's 1900 French text. I salute Davis and Etheridge for their present definitive translation, augmented by their splendid commentaries.

What adds to Bachelier's fame as a theorist is that he wrote long before Norbert Wiener provided a coherent basis for differential probability space. And even longer before the Itô stochastic calculus became available to Robert Merton or Fischer Black.

Even more impressive to lay students of how modern science evolved is that Bachelier beat Albert Einstein to the punch in analyzing what is essentially Brownian motion. Einstein's 1905 physics was impeccable. But as pure mathematics, Bachelier had already overlapped Einstein's findings and beyond that explicated how Fourier's derivation of the partial differential equation of heat applied isomorphically to the diffusions of probabilities. The famous Fokker–Planck and Chapman–Kolmogorov equations could, therefore, also carry the name of Bachelier first.

Notions today of 'Wall Street as a random walk' did get an important boost from Bachelier. But earlier, as far back as 1930, Holbrook Working's prolific Stanford research on future prices had documented the similarities between random-number sequences and time profiles of actual wheat and stock prices. One must still acknowledge that doctrines of 'efficient markets' did have anticipations in Bachelier.

Early on, discoverers of Bachelier realized that Bachelier's strict text involved price changes subject to absolute Gaussian distribution. By contrast, limited-liability common stocks can both rise and fall, but none of their prices can, by definition, go negative. Therefore, opportunistically I suggested replacing Bachelier's absolute Gaussian distribution by 'geometric' Brownian motion based on log-Gaussian distributions. Independently, the astronomer M. F. M. Osborne made the same suggestion, based on some analogy with Weber–Fechner laws in historic psychology.

Novel notions in science naturally invoke opposition from the ruling savants of normal science. That is why Bachelier's mathematics and efficient-market claims met with resistance—resistance from diverse sources.

(1) If Wall Street was only a casino, then must not 'economic law' be denied? That was the accusation of the economists assigned to frame a vote of thanks for (Sir) Maurice Kendall's 1953 Royal Statistical Society lecture that reported quasi-zero serial (Pearsonian) correlations of price changes for future contracts on commodities, for common stocks, and for indexed portfolios of the above. In post-mortem chatter with my friends who had attended that lecture, I mischievously suggested:

> Work the other side of the street. Economic law in its purest form would expect rationally anticipated prices to bounce quasi-randomly to incoming shocks from unanticipated future events.

(Most of my own half-dozen expositions of market efficiency, based on martingales rather than white noise, definitely do accord with economic fundamentalism. Working, who was a pioneer in recognizing random components, nevertheless compared carry-overs and seasonal pricings for onions (a) when speculative markets were banned as illegal, and (b) when they were legalized. Under (b), runs of good and bad harvests resulted, respectively, in storage and carry-over patterns that matched closely the linear-programming paths generated by an omniscient Robinson Crusoe; without legal bourses, what later became known as 'intertemporal Pareto non-optimality' was the rule in the Hobbesian jungle.)

(2) Old-guard resistance to post-Bachelier finance is well exemplified by the distinguished Nobelist and libertarian Milton Friedman. Early in the 1950s he had reacted adversely to Harry Markowitz's paradigm of portfolio optimization by mean-variance quadratic programming. It was not economics; nor was it at all interesting mathematics. Nor was this a hasty, tentative diagnosis. Forty years later in interviews with Reuters and the Associated Press, this truly great economist opined that the names of Markowitz, Sharpè, and Merton Miller would not be on connoisseurs' lists of 100 likely Nobel candidates. In some university junior common rooms, this incident brought into remembrance Max Planck's methodological dictum: science progresses funeral by funeral.

(3) Decades ago it was still not uncommon for most economists to doubt the usefulness of highfalutin mathematics for a social science like economics. When they overheard palaver about, say, Merton–Itô stochastic calculus, they were 'agin' it'. A different reaction comes from the lively twenty-first century school of 'behavioral economists'. These are well aware that few Wall Street locals can compute whether independent tosses of two fair coins will be more likely to result in both heads and tails rather than in two heads or two tails. So, how can their markets become and stay 'efficient'?

Within the sect of behaviorists, there are sophisticates who develop models in which most persons do diverge from Bachelier-type behavior, and thereby they do open up for more subtle traders temporary profits garnered from correctly correcting aberrant pricing patterns.

(4) Of course the harshest critics of efficient markets are all the brokers and investment bankers whose livelihoods would disappear in a

world of precisely efficient markets. Outnumbering these professionals as critics of market efficiency are the thousands of nongifted individuals who like a gamble but lack any flair at all for successful long-term risk-correct trading. They pursue and do not catch the fool's gold they seek.

Some very rare minds do have the special talent and flair needed for a good long-term batting average. More significant are the large universities, foundations, and millionaire family fortunes. Why do they outperform the noise traders and the small-college treasurers? I believe, but cannot prove, that it is largely because big money can (legally!) learn early more future-relevant information. Of course, those with earlier correct, not-yet-discounted, information do possess a Maxwell Demon that (joke) can defy the second law of thermodynamics.

(5) Perhaps methodologically most interesting—and certainly just that to the great Henri Poincaré, one of Bachelier's mentors—is the basic puzzle of how deductive mathematics usefully illuminates empirical behavior. The Bachelier tribe filters past oscillatory data, to nominate plausible profitable risk-corrected bets and hedges. These are guaranteed to perform well when applied to ideal *stationary* time series.

Real life, in Wall Street or Lasalle Street, or the City of London, never accords perfectly to a stationary time series. Yet, should traders deny *any* stationarity to the economic history record—past, present, or future? A true incident will explicate these points.

Some years ago there was a call option on a certain stock; I will call it Federal Finance. Traders applied to it standard Black–Scholes procedures, utilizing its past (and presumably extrapolatable) 'volatility' parameter. Ribbon clerks and assistant professors at MIT then noticed that, taking account of this call's exercise price, it was being grossly underpriced. So they bought some calls. Instead of their acumen being rewarded, the call went further down in market price. How could that be? A Black–Scholes hedge is a perfect hedge?

Well, there is no perfect hedge. Every Black–Scholes formula needs to have the right numerical volatility factor put in it. Who can ever know such perfection with certainty? The true tale tells its own lesson. Legally, but secretly, Federal Finance had contracted to be taken over at an already determined price. That meant its true volatility parameter became zero, far below past recorded volatilities. All money anted up was lost. (Moral: the crucial moment for intuitive traders is the one in which they compel themselves to disbelieve their own clever model. Time to go fishing, rather than risk more good money going bad. Long Term Capital Management seemingly never reached that crucial moment?)

We all owe gratitude to Professors Davis and Etheridge, whose labors and erudition have provided illuminating homage to a long underrated science hero. *Bon appetit!*

Paul A. Samuelson
June 2006

Preface

On 29 March 1900 Louis Bachelier defended his doctoral thesis, entitled 'Théorie de la spéculation', before an august jury of Parisian mathematicians. From their point of view this was a distinctly odd topic; in their words 'far away from those usually treated by our candidates'. Their report—reprinted below—was, however, couched in generous terms and rightly credits Bachelier with a high degree of originality. In fact, his originality was greater than they realized. His achievement was to introduce, starting from scratch, many of the concepts and ideas of what is now known as stochastic analysis, including notions generally associated with the names of other mathematicians working at considerably later dates. He defined Brownian motion—predating Einstein by five years— and the Markov property, derived the Chapman–Kolmogorov equation and established the connection between Brownian motion and the heat equation. The purpose of all this was to give a theory for the valuation of financial options. Bachelier came up with a formula which, given his mathematical model of asset prices, is correct—though not for the reason he thought it was. It was seventy-three years before Fischer Black and Myron Scholes produced, in the first decisive advance since 1900, their eponymous option pricing formula, very similar to Bachelier's but right for the right reasons.

Beyond providing the reader with an accurate and accessible translation of Bachelier's original thesis, together with access to the thesis itself, the purpose of this book is to trace the twin-track intellectual history of stochastic analysis on the one hand and financial economics on the other. These two subjects, united in Bachelier's work but then immediately separated, only coalesced again more than fifty years later, when a postcard dropped onto the desk of economist Paul Samuelson advising him to read Bachelier. The results were spectacular. Within twenty years the Nobel Prize-winning theory of option pricing and hedging had been developed by Black, Scholes and Robert Merton. After a further ten years the whole theory of 'arbitrage pricing' had been worked out and there was a burgeoning industry of trading of options on stocks, indices, foreign exchange, interest rates and commodities. Today this industry has grown to massive proportions, and expanded from 'plain vanilla' options to include trading in a whole range of 'exotic' contracts, as well as in further asset classes such as credit risk. None of this could have happened without the theories we are about to describe.

The story is a curious one. Stochastic analysis was developed over the course of the twentieth century through the work of a distinguished group of physicists and mathematicians including Einstein, Wiener, Kolmogorov, Lévy, Doob, Itô and Meyer. All of these except Einstein were aware of Bachelier, but he was for most of the time a marginal figure, and certainly nobody paid the slightest attention to the economic side of his work. Among economists, Bachelier was simply forgotten. Yet when he was rediscovered in the 1950s it was found that not only was his economics very close to the mark, but also stochastic analysis in its now highly developed form was exactly the tool needed to finish the job. No goal-oriented research programme could have done better.

The core of this book is a new English translation of 'Théorie de la spéculation'. We hope the reader will enjoy Bachelier's somewhat quirky style but also the quality of his intuitive grasp and—where he really was ahead of his time—his use of arguments based on sample path properties to calculate probabilities. To introduce the thesis we give, in Chapter 1, a discussion of options and their uses and some description of Bachelier's intellectual milieu in Paris at the turn of the twentieth century. We also describe the contemporary market in *les rentes*, the French government bonds which were the object of Bachelier's study. A translation of the report on the thesis, signed by Appell, Poincaré and Boussinesq, is included at the end of the thesis translation.

The final chapter traces the development of probability and financial economics from 1900 to 1981. Stochastic analysis is followed from the early work on Brownian motion of Bachelier and Einstein to Meyer's *théorie générale des processus* of the 1960s and 1970s, which in a very definite way completed the theory of stochastic integration. In the financial economics of options, nothing of note happened between Bachelier's thesis and its rediscovery by Samuelson. We describe the steps that led from this rediscovery to the option pricing theory of Black, Scholes and Merton and finally to a complete theory of arbitrage pricing through the work of Kreps, Harrison and Pliska. It is generally agreed that their 1979 and 1981 papers complete the conceptual framework of the subject (without, as they would be the first to agree, answering every technical question!).

One of us (MD) had the privilege of discussing the Bachelier saga personally with Paul Samuelson in a meeting at the Massachusetts Institute of Technology (MIT) in late 2003. We are honoured and delighted that Professor Samuelson has been able to contribute a foreword to this book. Several other participants in the story have been kind enough to provide us with very helpful information, advice and comments; we would particularly like to thank Robert Merton and Joseph Doob (who sadly died in 2004). More generally, our debt is to the lively community in mathematical finance—in academia and the markets—and we hope that people in that community and beyond will take as much pleasure as we have in stepping back a bit to trace how the subject came to be the way it is today.

Mark Davis and Alison Etheridge

Mathematics and Finance

'Janice! D'ya think you can find that postcard?'

Professor Paul A. Samuelson was in his office at MIT in the Autumn of 2003 relating how, several decades earlier, he had come across the PhD thesis, dating back to 1900, in which Louis Bachelier had developed a theory of option pricing, a topic that was beginning to occupy Samuelson and other economists in the 1950s. Although no economist at the time had ever heard of Bachelier, he was known in mathematical circles for having independently invented Brownian motion and proved some results about it that appeared in contemporary texts such as J. L. Doob's famous book *Stochastic Processes*, published in 1953. William Feller, whose influential two-volume treatise *An Introduction to Probability Theory and Its Applications* is widely regarded as a masterpiece of twentieth century mathematics, even suggested the alternative name Wiener–Bachelier process for the mathematical process we now know as Brownian motion. The story goes that L. J. ('Jimmie') Savage, doyen of mathematical statisticians of the post-World War II era, knew of Bachelier's work and, with proselytizing zeal, thought that the economists ought to be told. So he sent postcards to his economist friends warning them that if they had not read Bachelier it was about time they did. Hence Samuelson's appeal to Janice Murray, personal assistant *extraordinaire* to the Emeritus Professors at MIT's Department of Economics.

'When do you think you received it?' she enquired. 'Oh, I don't know. Maybe thirty-five years ago.'

How was Janice going to handle a request like that? For one thing, his timing was way off: it was more like forty-five years. But Janice Murray did not get where she is today without diplomatic skills.

'There's one place it might be', she said, 'and if it isn't there, I'm afraid I can't help you.'

It wasn't there.

Jimmie Savage's postcards—at least, the one sent to Samuelson—had spectacular consequences. An intensive period of development in financial economics followed, first at MIT and soon afterwards in many other places as well, leading to the Nobel Prize-winning solution of the option pricing problem by Fischer Black, Myron Scholes and Robert Merton in 1973. In the same year, the world's first listed options exchange opened its doors in Chicago. Within a decade, option trading had mushroomed into a multibillion dollar industry. Expansion, both in the volume and the range of contracts traded, has continued, and trading of option contracts is firmly established as an essential component of the global financial system.

In this book we want to give the reader the opportunity to trace the developments in, and interrelations between, mathematics and economics that lay behind the results and the markets we see today. It is indeed a curious story. We have already alluded to the fact that Bachelier's work attracted little attention in either economics or business and had certainly been completely forgotten fifty years afterwards. On the mathematical side, things were very different: Bachelier was not at all lost sight of. He continued to publish articles and books in probability theory and held academic positions in France up to his retirement from the University of Besançon in 1937. He was personally known to other probabilists in France and his work was cited in some of the most influential papers of the twentieth century, including Kolmogorov's famous paper of 1931, possibly the most influential of them all.

Bachelier's achievement in his thesis was to introduce, starting from scratch, much of the panoply of modern stochastic analysis, including many concepts generally associated with the names of other people working at considerably later dates. He defined Brownian motion and the Markov property, derived the Chapman–Kolmogorov equation and established the connection between Brownian motion and the heat equation. Much of the agenda for probability theory in the succeeding sixty years was concerned precisely with putting all these ideas on a rigorous footing.

Did Samuelson and his colleagues really need Bachelier? Yes and no. In terms of the actual mathematical content of Bachelier's thesis, the answer is certainly no. All of it had subsequently been put in much better shape and there was no reason to revisit Bachelier's somewhat idiosyncratic treatment. The parts of the subject that really did turn out to be germane to the financial economists—the theory of martingales and stochastic integrals—were in any case later developments. The intriguing point here is that these later developments, which did (unconsciously) to some extent follow on from Bachelier's original programme, were made almost entirely from a pure mathematical perspective, and if their authors did have any possible extra-mathematical application in mind—which most of them did not—it was certainly not finance. Yet when the connection was made in the 1960s between financial economics and the stochastic analysis of the day, it was found that the latter was so perfectly tuned to the needs of the former that no goal-oriented research programme could possibly have done better.

In spite of this, Paul Samuelson's own answer to the above question is an unequivocal 'yes'. Asked what impact Bachelier had had on him when he followed Savage's advice and read the thesis, he replied 'it was the tools'. Bachelier had attacked the option pricing problem—and come up with a formula extremely close to the Black–Scholes formula of seventy years later—using the methods of what was later called stochastic analysis. He represented prices as stochastic processes and computed the quantities of interest by exploiting the connection between these processes and partial differential equations. He based his argument on a martingale assumption, which he justified on economic grounds. Samuelson immediately recognized that this was the way to go. And the tools were in much better shape than those available to Bachelier.

From an early twenty-first century perspective it is perhaps hard to appreciate that an approach based on stochastic methods was a revolutionary step. It goes back to the question of what financial economists consider to be their business. In the past this was exclusively the study of financial markets as part of an economic system: how they arise, what their role in the system is and, crucially, what determines the formation of prices. The classic example is the isolated island economy where grain-growing farmers on different parts of the island experience different weather conditions. Everybody can be better off if some medium of exchange is set up whereby grain can be transferred from north to south when there is drought in the south, in exchange for a claim by northerners on southern grain which can be exercised when weather conditions in the south improve. In a market of this sort, prices will ultimately be determined by the preferences of the farmers (how much value they put on additional consumption) and by the weather. If one wants a stochastic model of the prices, one should start by modelling the participants' preferences, the weather and the rules under which the market operates. To take a purely econometric approach, i.e. present the prices in terms of some parametric family of stochastic processes and estimate

CHAPTER 1

the parameters using statistical techniques, is to abandon any attempt at understanding the fundamentals of the market. Understandably, any such idea was anathema to right-thinking economists.

When considering option pricing problems, however, the situation is fundamentally different. If the price of a financial asset at time t is S_t, then the value of a call option on that asset exercised at time T with *strike K* is $H_T = \max(S_T - K, 0)$ so that H_T is a deterministic function of S_T.[1] This is why call options are described as *derivative securities*. The option pricing problem is not to explain why the price S_T is what it is, but simply to explain what is the relationship between the price of an asset and the price of a derivative security written on that asset. As Bachelier saw, and Black and Scholes conclusively established, this question is best addressed starting from a stochastic process description of the 'underlying asset'. It is not bundled up with any explanation as to why the underlying asset process takes the form it does. In fact, Bachelier did have the right approach, although not the complete answer, to the option pricing problem—and at least as good an answer as anyone for fifty years afterwards—and his service to posterity was to point Samuelson and others in the right direction at a time when the mathematical tools needed for a complete solution were lying there waiting to be used.

Since it was not the norm at the time to include full references, it is impossible to know how much of the literature he was familiar with, but Bachelier's work did not appear from an economic void. Abstract market models were already gaining importance. Starting in the middle of the nineteenth century several attempts had been made to construct a theory of stock prices. Notably, Bachelier's development mirrors that of Jules Regnault, who, in 1853, presented a study of stock market variations. In the absence of new information which would influence the 'true price' of the stock, he believed that price fluctuations were driven by transactions on the exchange which were in turn driven by investors' expectations. He likened speculation on the exchange to a game of dice, arguing that future price movements do not depend on those in the past and there are just two possible outcomes: an increase in price or a decrease in price, each with probability one-half. (These probabilities are *subjective* probabilities arising from incomplete information, and different assessments of that information, on the part of the market players.) Regnault's study of the relationship between time spans and price variations led him to his *law of differences* (loi des écarts) or square root law: the spread of the

---

[1] A fuller description of options is given below.

prices is in direct proportion to the square root of the time spans. Bachelier provides a mathematical derivation of this law which governs what he calls the 'coefficient of instability', but Regnault was no mathematician and his theoretical justification is unconvincing. He represented the *true price* of a security during an interval as the centre of a circle with the interior of the circle representing all possible prices. The area of the circle grows linearly with time and so the deviations from the true price grow with the radius of the circle, which is the square root of time. Regnault did, on the other hand, produce a convincing *verification* of his law, based on price data that he had compiled on the 3% *rentes*[2] from 1825 to 1862.

Of course Bachelier was concerned not just with stocks, but also with the valuation of derivative securities. From the beginning of the nineteenth century, it was common to value stocks relative to a fixed bond. Instead of looking at the absolute values of the stocks, tables were compiled that compared their relative price differences with the chosen bond and grouped them according to the size of the fluctuations in these differences. In the same way options were analysed relative to the underlying security and, in 1870, Henri Lefèvre, former private secretary to Baron James de Rothschild, developed the geometric representation of option transactions employed by Bachelier thirty years later. Lefèvre even used this visual approach to develop 'the abacus of the speculator', a wooden board with moveable letters which investors could reposition to find the outcome of a decision on each type of option contract. This ingenious invention was similar to the *autocompteur*, a device that he had previously introduced for computing bets on racehorses.[3]

Bachelier's thesis begins with a detailed description of some of the derivative contracts available on the exchange and an explanation of how they operate. This is followed by their geometric representation. Next comes the random walk model. Once this model is in place, economics takes a back seat while he develops a remarkable body of original mathematics. Assuming only that the price evolves as a continuous, memoryless process, homogeneous in time and space, he establishes what we now call the Chapman–Kolmogorov equation and deduces that the distribution of the price at a fixed time is Gaussian. He then considers the probability of different prices as a function of time and establishes the

---

[2]Perpetual government bonds. They are described in detail below.

[3]The *autocompteur* is described in a little more detail in Preda (2004), where some more detail of the contributions of Regnault and Lefèvre can be found. See also Taqqu (2001).

square root law. He presents an alternative derivation of this by considering the price process as a limit of random walks. The next step is the connection between the transition probabilities and Fourier's heat equation, neatly translating Fourier's law of heat flow into an analogous law of 'probability flow'. There follow many pages of calculations of option prices under this model with comparisons to published prices. The final striking piece of mathematics is the calculation of the probability that the price will exceed a given level in a particular time interval. The calculation uses the reflection principle, well known in the combinatorial setting as Bachelier himself points out, but his direct proof of this result is a thoroughly modern treatment with paths of the price process as the basic object of study. The two things in the thesis that stand out mathematically are the introduction of continuous stochastic processes and the concentration on their *paths* rather than their value at a fixed time as the fundamental object of study. There is sometimes a lack of rigour, but never a shortage of originality or sound intuition.

OPTIONS AND RENTES

Option contracts have been traded for centuries. It is salutary to realize how sophisticated the financial markets were long ago. Richard Dale's book *The First Crash* describes the London market of the late seventeenth and early eighteenth centuries, where forward contracts, put options (called 'puts') and call options (called 'refusals') were actively traded in Exchange Alley. It seems that the British were at the time a nation of inveterate gamblers. One could bet on all kinds of things: for example, one could buy annuities on the lives of third parties such as the Prince of Wales or the Pretender. Had they known, these luminaries might have taken comfort from the idea that a section of the population had a direct stake in their continued existence, but they would have been less pleased to discover that an equal and opposite section had a direct stake in their immediate demise.

Like these annuities, options were simply a bet, and a dangerous one at that because of the huge amount of leverage involved. Think of a *one-year at-the-money call option* on a stock[4], for which the premium is 10% of the current stock price; thus a £1 investment buys options on £10 of stock. If the stock price fails to rise, the investment is simply lost. On

---

[4]The strike price is the current price S_0 and so the exercise value is $\max(S_1 - S_0, 0)$, where S_1 is the price in one year's time. Here $\max(a, b)$ denotes the greater of two numbers a, b.

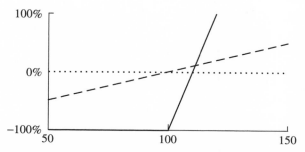

Figure 1.1. Returns from investing in asset (dashed line) and investing in call option (solid line) as functions of asset price at maturity date. Initial asset price is 100.

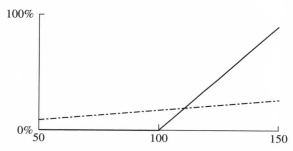

Figure 1.2. Returns from investing in option strategy with money-back guarantee return (solid line) and simple investment strategy with money-back guarantee.

the other hand if it rises by 20% then the option pays £2, a 100% return on the investment (as opposed to the 20% return gained by investing in the stock itself). The option investor is taking a massive risk—the risk of losing his entire investment—to back his view that the price will rise. For this reason option contracts have always had a slightly disreputable air about them, which continues to the present day. Lawsuits are regularly taken out by aggrieved parties claiming that the risks in option-like investments were not properly explained to them.

Figure 1.1 shows the return as a function of price for simple investment and for investment in options. Viewed in this way, a 'naked' position in options does seem like playing with fire. Nonetheless, by mixing with other investments, the buyer of an option can take advantage of leveraged returns while limiting his downside loss liabilities. For example, investment products are sometimes offered that guarantee investors at least their 'money back' after, say, five years. Suppose the interest rate for a five-year deposit is 4%. Then a deposit of £82.19 matures with a value of £100 in five years' time. The investment company can offer the money-back guarantee by investing £82.19 of each £100 in this way and 'playing the market' with the remaining £17.81. Figure 1.2 shows the return as a function of underlying asset for investment in the asset itself or in options as described above. Arguably, the latter provides a more attractive return profile to investors: there is the possibility of a substantial gain and the downside is limited to interest lost by not just keeping the money in the bank.

A more benign, and economically far more important, use of options is in connection with *hedging* risks. Used in this way, an option is not a leveraged investment but, on the contrary, is expressly designed to offset risk. This effect is closely related to a simpler transaction, the forward contract. Suppose Alice in London wants to buy, off-plan, a flat on the Costa del Sol. The agreed price is €600,000, to be paid when the building is completed in one year's time. This is equal to £400,000 at today's exchange rate of €1.5 : £1. But of course Alice now has exchange rate risk: if the rate of exchange were to fall to 1.4 : 1, a not grossly improbable contingency, Alice's bill rises by an unpleasant £28,571. She can protect herself against this in the following way. Let us say that Alice is able to borrow in England at an annual rate of 6%, and has a deposit account in Spain paying 2%. If she borrows £600,000/(1.5 × 1.02) = £392,157, converts it into euros and deposits this sum in the Spanish account, the value of the account in one year will be exactly the €600,000 she has agreed to pay. But meanwhile her negative balance in England will be £392,157 × 1.06 = £415,686, so the effective exchange rate for the transaction is 600,000/415,686 = 1.443, which is the rate she should be allowing for when agreeing the price. But, in addition to protecting herself against a loss, Alice has also protected herself against a profit: if the foreign exchange rate were instead to move to 1.6, Alice would, in retrospect, have been better off sitting back and taking the profit of £25,000. What she wants is the best of both worlds: a fixed exchange rate X and the right to use either that rate or the market spot rate, whichever is the more favourable. This is the classic call option. By buying a call option, Alice is in effect insuring herself against the exchange rate falling below X, so she should pay an insurance premium for that. What that premium should be is the option pricing problem.

The options that Bachelier was concerned with were written on *les rentes*, perpetual French government bonds. City or state government bonds have a long history, intricately bound up with the need to raise money to wage war and to pay reparations after defeat.[5] In the earliest examples the buyer was given little choice in the matter, but in 1522 perpetual annuities called *rentes* were floated free of coercion, secured by the tax on wine. Under these contracts the state undertook to pay an annuity for an indefinite period. They did not repay the principal although they did reserve the right to redeem the contract at any time. It did not take the *rentiers* long to appreciate the benefits of having taxes imposed on their behalf and they soon offered additional loans. But *rentes* have a chequered history. Twice in the seventeenth century payments were simply suspended. Finance ministers disallowed *rentes* created by their predecessors and created new ones, there was no uniformity in conditions and no clear record of the outstanding government obligation. Each war saw more issues so that, for example, between 1774 and 1789 the total government debt tripled. The bad credit of the revolutionary government led them to issue *assignats*, obligations supposedly backed by land seized from nobility and the church, which bore interest and were to be redeemed in five years. But the interest was reduced and then abolished and the *assignats* were issued in smaller and smaller denominations so that they eventually became no more than paper money which lost all value if they were not redeemed. Finally, in 1797, they were declared valueless. In the same year the two-thirds bankruptcy law was passed whereby only one-third of the interest on the national debt and one-third of pensions was paid in cash, with the balance in land warrants of little or no real value. All confidence in the system was lost.

It was Napoleon who reformed the French finances. Starting in 1797 he established a budget, increased taxes and forcibly refunded the national debt through an issue of 5% *rentes*. All valid loans and titles to *rentes* were recorded in the *Grand Livre*, the great book of public debt, which had been created in 1793, and France entered the modern era of uniform *rentes*. To service the debt, Napoleon reestablished the *caisse d'amortissement*, which he deposited, in return for shares, in the Banque de France. This semi-private organization, created by a group of his supporters in 1800, was not only granted the monopoly on Parisian banknotes, it also managed the *rentes*.

The nineteenth century was a period of political turmoil in France, nonetheless there was a large growth in banking and increasingly orderly finance. The *rentes* recovered quickly from successive political crises

---

[5]For a more detailed history, see Homer and Sylla (2005).

and although the interest was progressively reduced during the course of the century, by 1900 they were very popular with French investors as a relatively secure investment. Once issued their prices fluctuated, but the *rentes* had a nominal value, typically Fr 100, and paid a fixed return usually between 3% and 5%. At the time when Bachelier wrote his thesis, the nominal value of the debt was 26 billion francs (Taqqu 2001) and there was very considerable trade in *rentes* on the Paris Bourse, where they could be sold for cash or as forwards or options.

In 1914 stock exchanges all over the world were forced to close and whilst this did not signal the death of the international bond market, the collapse of the French franc over the course of the century, losing 99% of its dollar value in the period 1900–1990, spelt the end of an era for the *rentiers* (Ferguson 2001).

GAMBLING STRATEGIES AND MARTINGALES

In the introduction to his 1968 book *Probability*, Leo Breiman points out that probability theory as we know it today derives from two sources: the mathematics of measure theory on the one hand, and gambling on the other.[6] Perhaps its unsavoury relationship with games of chance is at least partially responsible for the fact that probability took a long time to be regarded as a respectable branch of mathematics. The flavour is caught to perfection by William Makepeace Thackeray in Chapter 64 of *Vanity Fair*:

> There is no town of any mark in Europe but it has its little colony of English raffs—men whose names Mr. Hemp the officer reads out periodically at the Sheriffs' Court—young gentlemen of very good family often, only that the latter disowns them; frequenters of billiard-rooms and estaminets, patrons of foreign races and gaming-tables. They people the debtors' prisons—they drink and swagger—they fight and brawl— they run away without paying—they have duels with French and German officers—they cheat Mr. Spooney at *écarté*—they get the money and drive off to Baden in magnificent britzkas—they try their infallible martingale and lurk about the tables with empty pockets, shabby bullies, penniless bucks...

The 'infallible martingale' is a strategy for making a sure profit on games such as roulette in which one makes a sequence of independent bets. The strategy is to stake £1 (on, say, a specific number in roulette)

---

[6]He credits Michel Loève and David Blackwell, respectively, for teaching him the two sides.

Figure 1.3. Harness with martingale.

and keep doubling the stake until that number wins. When it does, all previous losses and more are recouped and one leaves the table with a profit. It does not matter how unfavourable the odds are, only that a winning play comes up eventually. In his memoirs, Casanova recounts winning a fortune at the roulette table playing a martingale, only to lose it a few days later.

The word 'martingale' has several uses outside gambling. It can mean a strap attached to a fencer's épée, or a strut under the bowsprit of a sailing boat, but the most common usage is equestrian: the martingale refers to the strap of a horse's harness that connects the girth to the noseband and prevents the horse from throwing back its head (see Figure 1.3). Like the gambling strategy, it allows free movement in one direction while preventing movement in the other. The mathematician Paul Halmos once sent J. L. Doob, who did more than any other single mathematician to develop the mathematical theory of martingales, an equestrian martingale. Doob had no idea what it was or why he had been sent it (Snell 2005).

The martingale is not infallible, as the penniless bucks whose names Mr Hemp read out at the Sheriffs' Court could attest. Nailing down why, in precise terms, had to await the development of the theory of martingales (in the mathematical sense) by Doob in the 1940s. The term was introduced into probability theory by Ville in 1939, who initiated its use to describe the fortune of a player in a fair game rather than the gambling strategy employed by that player. A martingale is then a stochastic processes X_t such that the expected value of the process at some future time, given its past history up to today, is equal to today's value. We write this $X_s = \mathbb{E}[X_t \mid \mathcal{F}_s], t > s$. Roulette is not a fair game: the player's fortune is a *supermartingale*, meaning that the expected future value is *less than* today's value, $X_s \geqslant \mathbb{E}[X_t \mid \mathcal{F}_s]$. One of Doob's key results is the *optional sampling theorem*. A *stopping time* (or *optional time* in Doob's parlance) is a random time whose occurrence by time t can be detected by observing the evolution of the process X_s for $s \leqslant t$. For example, the time of the first winning play in roulette is a stopping time. The optional sampling theorem shows in particular that if X_t is a bounded supermartingale (i.e. $|X_t| \leqslant c$ for some constant c) and S, T are two stopping times such that $S \leqslant T$ with probability 1, then $X_S \geqslant \mathbb{E}[X_T \mid \mathcal{F}_S]$, i.e. the supermartingale inequality continues to hold if fixed times s, t are replaced by stopping times S, T. For a bounded martingale, $X_S = \mathbb{E}[X_T \mid \mathcal{F}_S]$.

Now suppose that X_t is the player's fortune when he plays the martingale strategy at roulette and T is the time of the first winning play. X_t is a supermartingale since the odds are biased in favour of the bank. The conditions of the optional sampling theorem are not met since X_t is not bounded (losses double up until the first winning play occurs, but we do not know how long we have to wait for this). And indeed the conclusion of the theorem does not hold either: by definition $X_T > X_0$, so $\mathbb{E}[X_T] > \mathbb{E}[X_0]$. Suppose, however, that there is a *house limit*: the player has to stop if his accumulated losses ever reach some prescribed level K. The conditions of the optional sampling theorem are satisfied for the process[7] $Y_t = X_{t \wedge R}$, the player's fortune up to the point R where he is obliged to quit, so $\mathbb{E}[X_{T \wedge R}] \leqslant X_0$. But this inequality can only hold if there is a positive probability that $R < T$, that is, there must be a chance that the house limit is reached before the winning play occurs. Thus any house limit, however large, turns the 'martingale' into an unfavourable strategy in which the player may lose his shirt. Every house has a limit of some kind.

The martingale idea plays a big part in Bachelier's analysis, although he does not define it in any formal way and, of course, the name itself did not come into mathematical currency for a further thirty-nine years. Bachelier's dictum (perhaps inherited from Regnault) was (see p. 28) 'L'espérance mathématique du spéculateur est nulle' ('the speculator's expected return is zero'). The argument for this is based on market symmetry: any trade has two parties, a buyer and a seller, and they must agree on a price. It follows that there cannot be any consistent bias in favour of one or the other, so today's price must be equal to the expected value of the price at any date in the future: exactly the martingale property. In Bachelier's day, the option premium was paid in the form of a

---

[7] $t \wedge s$ denotes the lesser of s and t.

'forfeit' paid at the exercise time of the option, and only paid if it was not exercised. Bachelier's main pricing formula is obtained by taking the price process as scaled Brownian motion and computing the value of the forfeit such that the whole transaction has zero value. Given Bachelier's price model this answer is actually correct but not, as we shall see, for quite the reasons Bachelier thought it was.

Deciding when to quit a game is a very simple kind of gambling (or investment) strategy. A correct theory of option pricing requires consideration of more sophisticated strategies in which funds are switched between different traded assets in a quite complicated way. Suppose there are N traded assets[8] with price processes $S_t = (S_t^{(1)}, \ldots, S_t^{(N)})$. A trading strategy is an N-vector process θ with the interpretation that the ith component $\theta_t^{(i)}$ is the number of units of asset i held at time t. An obvious requirement is that θ_t must be non-anticipative, i.e. can depend only on market variables that have been observed up to time t. We say that θ_t is '$\mathcal{F}_t$-adapted', or just 'adapted', where $\mathcal{F}_t$ denotes the history of the market up to time t. We call θ_t a *simple* trading strategy on the time interval $[0, T]$ if there is a sequence of fixed or stopping times $0 = \tau_0 < \tau_1 < \cdots < \tau_m \leqslant T$ such that $\theta_t = \theta_{\tau_i}$ for $t \in [\tau_i, \tau_{i+1})$, that is, trades are executed only at a finite number of times τ_i. Let us write $\eta_i = \theta_{\tau_i}$. The strategy is *self-financing* if $\eta_i \cdot S_{\tau_{i+1}} = \eta_{i+1} \cdot S_{\tau_{i+1}}$, which just says that the trade at time τ_{i+1} only rearranges the investment portfolio, it does not change its total value. It is a matter of algebra to verify that the following equality holds for any simple self-financing strategy,

$$\theta_T \cdot S_T - \theta_0 \cdot S_0 = \int_0^T \theta_t \cdot dS_t, \qquad (1.1)$$

where the right-hand side denotes the sum suggested by the notation

$$\int_0^T \theta_t \cdot dS_t = \sum_{i=1}^N \int_0^T \theta_t^{(i)} dS_t^{(i)}$$
$$= \sum_{i=1}^N \left\{ \eta_m^i (S_\tau^i - S_{\tau_m}^i) + \sum_{k=0}^{m-1} \eta_k^i (S_{\tau_{k+1}}^{(i)} - S_{\tau_k}^{(i)}) \right\}. \qquad (1.2)$$

Equation (1.1) states that the change in portfolio value (on the left-hand side) is equal to the 'gain from trade' (on the right-hand side). We can use (1.1) as the definition of 'self-financing', a definition which will go

---

[8] In the classic Black–Scholes set-up, $N = 2$: $S_t^{(1)} = S_t$ is the price of the asset on which the option is written and $S_t^{(2)} = e^{rt}$ is a money-market account paying continuously compounding interest at rate r.

13

beyond the case of simple strategies. Black–Scholes style option pricing is largely concerned with finding self-financing trading strategies θ_t such that the final portfolio value $\theta_T \cdot S_T$ is equal to some pre-specified random variable, to wit, the option exercise value. Invariably, this cannot be achieved using simple strategies and we have to consider more general strategies which are in some sense limits of simple ones. The process of constructing these strategies and defining the corresponding gains from trade is exactly the process of constructing stochastic integrals with respect to martingales and semimartingales which was begun by Itô in the 1940s and completed by Meyer and Dellacherie in the late 1970s. As Itô explains in the foreword to the volume of his selected papers published in 1986, he noticed that, starting from any given instant, a Markovian process would perform a time homogeneous differential process for the infinitesimal future. This led him to the notion of a stochastic differential equation governing the paths of the particle which could then be made mathematically rigorous by writing it in integral form, but this intuition requires that when defining the integral of a simple integrand, the integrand should be evaluated at the left-hand end point of the interval, as it is in equation (1.2) above. In equation (1.2) we arrived at this definition from the natural economic requirement that the investment must be decided on *before* subsequent price moves are revealed. Thus Itô–Meyer–Dellacherie stochastic integrals are exactly the ones required for economic applications.

In J. L. Doob's obituary of William Feller in 1970 he writes:

> Mathematicians could manipulate equations inspired by events and expectations long before these concepts were formalized mathematically as measurable sets and integrals. But deeper and subtler investigations had to wait until the blessing and curse of direct physical significance had been replaced by the bleak reliability of abstract mathematics.

The deep and subtle mathematics that underlies today's financial markets is all-pervasive in engineering and the sciences. It has long since emerged from the wilderness of bleak reliability and is once again blessed and cursed.

And now to Bachelier. The next chapter is a translation of Louis Bachelier's thesis. We have endeavoured to reflect his written style. We have used two types of annotation: comment boxes to explain the financial contracts, and traditional footnotes for clarifications, corrections and historical comments. Bachelier's own footnotes are unnumbered and in italics. The original thesis itself is reproduced in facsimile in Chapter 4.

Théorie de la Spéculation

INTRODUCTION

The influences that determine the movements of the Exchange are innumerable; past, current and even anticipated events that often have no obvious connection with its changes have repercussions for the price. Alongside the, as it were, natural variations, artificial causes also intervene: the Exchange reacts to itself and the current movement is a function not only of previous movements but also of the current state. The determination of these movements depends upon an infinite number of factors; it is thus impossible to hope for mathematical predictability. Contradictory opinions about these variations are so divided that at the same time the buyers believe in an increase and the sellers in a decrease. The calculus of probabilities can no doubt never apply to movements of stock exchange quotations[1] and the dynamics of the Exchange will never be an exact science, but it is possible to study mathematically the static state of the market at a given instant, that is to say, to establish the probability distribution of the variations in price that the market admits at that instant. Although the market does not predict the movements, it does consider them as being more or less likely, and this probability can be evaluated mathematically. Research into a formula that expresses this probability does not appear to have been published until now; it will be the object of this work. I have deemed it necessary to first of all recall some theoretical notions relating to operations of the exchange adding certain new observations indispensable in our later research.

---

[1] Here, perhaps following Regnault (1853), Bachelier is distinguishing between 'gross price movements', that is, movements in an underlying 'true price' due to some external impetus, and fluctuations about the true price. It is the latter that he goes on to model.

CHAPTER 2

The Transactions of the Exchange

Transactions of the Exchange

There are two sorts of forward-dated transaction:

- forwards,[2]
- options.

These transactions can be combined ad infinitum, especially as one often treats several kinds of option.

Forwards

Forwards are completely analogous to spot transactions but one only settles price differences at a time that is fixed in advance, called the *liquidation date*. It falls on the last day of each month. The price, established at the liquidation date, to which one tallies all the transactions of the month is the *clearing price*.

The buyer of a forward limits neither his gains nor his losses. He gains the difference between the buying price and the selling price if the sale is made above the buying price, he loses the difference if the sale is made below. There is a loss for the seller of a forward who repurchases at a higher price than the original sale, there is gain in the opposite case.

Contangoes

The spot buyer cashes his coupons and can keep his shares indefinitely. Since a forward contract expires at the liquidation date, the buyer of the settlement must, to conserve his position until the next liquidation date, pay the seller an indemnity called a *contango*.*

> The buyer of the forward contract acquires all the rights of ownership immediately, and in particular receives any interest payments on the bond, but payment for the bond is only made at the liquidation date. The seller receives neither the selling price nor the interest and so must be compensated by the payment of a *contango* or *forwardation*. The contango varies from month to month, capturing the market's belief about a rise/fall in prices.
>
> The French word for contango, *report*, has entered modern French slang as a word for an accumulator bet.

[2]A literal translation here would be 'fixed contracts', we have adopted the modern terminology.

For the complete definition of contangoes I refer to specialist works.

16

The contango is different at each liquidation date; on government bonds it averages Fr 0,18 per Fr 3,[3] but it can be higher or zero; it can even be negative, in which case one calls it a *backwardation*. In this case the seller pays an indemnity to the buyer.

On the day of the coupon payment, the buyer of the contract receives from the seller the value of the coupon. At the same time, the price of the security falls by an equal amount[4]; buyer and seller therefore find themselves in the same relative position immediately after this transaction as before.

One sees that although the buyer has the advantage of cashing the coupons, on the other hand he must in general pay the contangoes. The seller, by contrast, receives the contangoes but pays the coupons.

Government Bonds with Contangoes

On government bonds, the coupon of Fr 0,75 per quarter represents Fr 0,25 per month, while the contango is almost always less than Fr 0,20. The difference therefore favours the buyer. This suggests the idea of buying bonds and carrying them forward indefinitely. This transaction is known as *the government bond with contangoes*; later we study its probability of success.

Equivalent Prices

In order to give a clearer account of the mechanism of coupons and contangoes, let us abstract all the other causes of price variation.

Since every three months a coupon of Fr 0,75 is detached from a government bond representing the interest on the money invested by the buyer, logically the bond must increase by Fr 0,25 every month. To the price currently quoted corresponds a price which in thirty days will be Fr 0,25 higher, in a fortnight Fr 0,125 higher and so on. All these prices can be considered as *equivalent*.

---

[3]Although once issued their prices fluctuated, *rentes* had a nominal value, typically Fr 100, and paid a fixed return. The rates ranged between 3% and 5%. The contango of Fr 0,18 referred to here is on a 3% bond with nominal value Fr 100.

[4]Bond prices are now typically quoted 'net' of accumulated interest. That is, a bond purchased between interest payment dates will actually cost the quoted price plus the value of any accumulated interest. In Bachelier's France, the quoted price included accumulated interest so that the price behaved like that of a dividend-paying stock, jumping down by an amount equal to the coupon payment on the day that the coupon was detached.

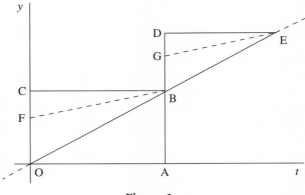

Figure 1.

The consideration of equivalent prices is much more complicated in the case of forward contracts. First of all it is evident that if the contango is zero, the forward must behave like the spot position and that the price must logically increase by Fr 0,25 per month.

Let us now consider the case when the contango is Fr 0,25. We take the x-axis to represent time (Figure 1), the length OA represents one month contained between two liquidation dates of which one corresponds to the point O and the other to A. The values on the y-axis represent prices.

If AB is equivalent to Fr 0,25, the logical movement of the spot price of the government bond will be represented by the straight line OBE.* Consider now the case when the contango is Fr 0,25. A little before the liquidation date, the spot price and the forward contract will be at the same price O, then the buyer of the forward must pay a contango of Fr 0,25. The price of the forward jumps swiftly from O to C and during the whole month follows the horizontal line CB. At B it coincides once again with the spot price and then increases in one jump of Fr 0,25 to D and so on.

In the case when the contango is a given quantity corresponding to the length OF, the price must follow the line FB, then GE and so on. The forward contract must therefore in this case, from one liquidation date to the next, grow by a quantity represented by FC, which one could call the *complement of the contango*.

All the prices from F to B on the line FB are *equivalent* at the different times to which they correspond.

---

*We suppose that there is no coupon payment during the interval under consideration, this does not change the proof.

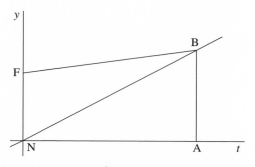

Figure 2.

In reality, the gap between the forward and the spot rate does not behave in a completely regular manner and FB is not a straight line, but the construction of equivalent prices that has just been made from the contango at the beginning of the month can be repeated at an arbitrary time represented in Figure 2 by point N. Let NA be the time that passes between the time N under consideration and the liquidation date represented by the point A.

During the time NA, the spot price must rise by the amount AB. Let NF be the gap between the spot price and the forward, all the prices corresponding to the line FB are *equivalent*.

True Price

The equivalent price corresponding to a given time is called the *true price* corresponding to that time. Knowledge of the true price is of very great importance; I am going to study how one determines it.

Let us denote by b the quantity by which the forward contract must logically increase in the interval of one day. The coefficient b generally varies little, its value each day can be exactly determined. Suppose that n days separate us from the liquidation date and let C be the gap between the price of the forward contract and the spot price. In n days, the spot price must increase by $\frac{25n}{30}$ centimes. The price of the forward contract, being bigger by the quantity C, must increase only by the quantity $\frac{25n}{30} - C$ during these n days, that is to say, by

$$\frac{1}{n}\left(\frac{25n}{30} - C\right) = \frac{1}{6n}(5n - 6C)$$

during one day. One therefore has

$$b = \frac{1}{6n}(5n - 6C).$$

19

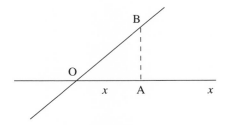

Figure 3.

The average over the last five years gives

$$b = 0,264 \text{ centimes.}$$

The true price of the forward contract corresponding to m days will be equal to the current quoted price augmented by the quantity mb.

Geometric Representation of Forward Contracts

A transaction can be represented geometrically in a very simple manner, the x-axis representing the different prices and the y-axis the corresponding profits.

Suppose that I had purchased a forward at the price represented by O in Figure 3, which I take to be the origin. At the price $x = $ OA, the transaction gives a profit x and, since the corresponding point on the y-axis must equal the profit, AB $=$ OA. The purchase of the forward is therefore represented by the line OB inclined at 45° to the line of prices.

The sale of a forward is represented in an inverse fashion.

Options

In a forward purchase or sale, buyers and sellers expose themselves to a theoretically unlimited gain or loss. In the options market, the buyer pays more for his certificate than in the forwards market, but his loss is limited in advance to a certain sum that is given by the amount of the premium. The seller of the option has the advantage of selling for more money, but he cannot gain more than the value of the premium.

One also has downside options[5] that limit the losses of the seller. In this case the transaction is made at a lower price than that of the forwards market. One only negotiates such downside options in speculation

---

[5] An alternative translation would be 'options at a discount'. In modern terminology this is just a European put option.

on commodities; in speculation on assets one obtains a downside option by selling a forward and simultaneously buying an option.[6]

To fix ideas, I will only occupy myself with upside options.

Suppose for example that 3% bonds cost Fr 104 at the start of the month: if we buy 3000 on a forward contract, we expose ourselves to a loss that becomes considerable if there is a sharp fall in bond prices. To avoid this risk, we can buy an option at 50 centimes* paying now, not Fr 104, but Fr 104,15, for example. It is true that our buying cost is higher, but our loss is limited, whatever the fall, to 50 centimes per Fr 3, that is to say, Fr 500.[7]

> This option would now be known as a *European call option*. The buyer of the option has the right, but not the obligation, to buy bonds with nominal value Fr 100 for Fr 104,15 at the end of the month. If he does not exercise that right, then he must pay a forfeit of 50 centimes.

The transaction is the same as if we had bought forwards at Fr 104,15, the price not being able to fall more than 50 centimes, that is to say fall below Fr 103,65.

The price of Fr 103,65 in the current case is the *exercise price of the option*.[8] One sees that the exercise price is equal to the price at which the option is negotiated minus the premium of the option.

Call Date of Options

The day before the liquidation date, that is, the day before the last day of the month, is the *call date of options*. Let us return to the previous example and suppose that at the call date the price of the bond is less than Fr 103,65; we will *abandon* our option, which will profit our seller. If, on the other hand, the price at the call date is higher than Fr 103,65 our transaction will be transformed into a forward contract: in this case one says that the option is *exercised*.

---

[6]This observation is called 'put–call parity'.

* *One says 'une prime dont' for 'une prime de' and uses the notation 104,15/50 to denote a transaction made at the price Fr 104,15 for 50 centimes.*

[7]Recall Bachelier's notation for *la rente* here, Fr 3 refers to a 3% bond with a nominal value of Fr 100. By buying 3000 Bachelier means 1000 such bonds with consequent annual coupon payments totalling Fr 3000.

[8]A literal translation would be 'foot of the option'. We have adopted the modern term *exercise price*. It is the minimum price at which the option will be exercised. Bachelier uses *price* to mean the exercise price plus the premium. In contrast to modern markets, in Bachelier's world, the premium would have been paid at the time of expiry of the option, not on purchase.

21

In summary, an option is exercised or abandoned according to whether the price at the call date is above or below the exercise price of the option.

We see that options do not run until the liquidation date: if at the call date the option is exercised, it becomes a forward and liquidates the next day. In all that follows, we suppose that the clearing price coincides with the price at the call date of options. This hypothesis can be justified, because nothing prevents liquidation of these transactions at the call date for options.

The Spread of Options

The spread[9] between the forward price and that negotiated in an option depends on a large number of factors and varies all the time. At the same time, the difference is bigger as the premium is smaller, for example the option at 50 centimes is obviously cheaper than the option at 25 centimes.

The spread of an option decreases more or less regularly from the beginning of the month until the call date, when this spread becomes very small, but, according to the circumstances, it can be very irregular and be bigger a few days before the call date than at the beginning of the month.

Options for the End of the Next Month

One deals not only with options expiring at the end of the current month but also with options expiring at the end of the next month. The spread of these is necessarily bigger than those of options for the current month but it is smaller than one would expect from taking the difference between the price of the option and that of the forward. It is necessary, in fact, to deduct from this apparent spread the size of the contango at the end of the current month. For example, the average spread of the option at 25 centimes at forty-five days before the call date is about 72 centimes, but as the average contango is 17 centimes, the spread is in reality only 55 centimes.

The detaching of a coupon lowers the price of the option by an amount equal to the value of the coupon. If, for example, on 2 September I buy an option at 25 centimes for Fr 104,50 to expire at the end of the current month,[10] the price of my option will have become Fr 103,75 on

---

[9]The spread of an option is the exercise price plus the premium (which is the price negotiated in the option) minus the forward price.

[10]In his previous notation Bachelier should have denoted this as 104,50/25. In fact

16 September after the coupon is detached. The exercise price of the option will be Fr 103,50.[11]

Options for the Next Day

One negotiates, especially outside the exchange,[12] options at 5 centimes and sometimes at 10 centimes for the next day. The call time for these little options is at 2 p.m. each day.

Options in General

In an options market for a given settlement date, there are two factors to consider: the size of the premium and the spread from the forward price. It is clear that the bigger the premium, the smaller the spread.

To simplify the negotiation of options, they have been classified into three types according to the three simplest hypotheses on the size of the premium and the spread:

1. The premium of the option is constant and its spread is variable. It is this sort of option that is negotiated on securities; for example on the 3% bond one negotiates options at 50 centimes, 25 centimes and 10 centimes.
2. The spread of the option is constant and its premium is variable. This applies to downside options (that is to say, the sale of a forward contract against the purchase of an option).
3. The spread of the option is variable as is its premium, but the two quantities are always equal. It is in this way that one negotiates options on commodities. It is clear that by employing this last system at a given moment one can only treat a single option for the same maturity.

Remark on Options

We shall examine the law that regulates the spreads of options. However, we can, right away, make a quite curious observation: the value of an option must become bigger as its spread becomes smaller, but this obvious fact is not enough to show that the use of options is rational. I have

---

there appears to be a misprint here. The numbers that follow make sense only if we replace 104,50/25 by 104,00/25.

[11]If the option is exercised then the holder will receive all the payments made on the security during the term of the option. After the interest payment, it therefore becomes worthwhile to exercise the option if the price is above Fr 103,50.

[12]A literal translation here would be 'behind the scenes'.

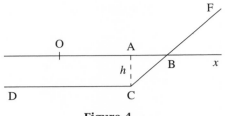

Figure 4.

in fact known for several years that it would be possible, while accepting this fact, to imagine transactions where one of the parties makes a profit at all prices. Without reproducing the calculations, which are easy but rather laborious, I content myself with presenting an example. The following transaction,

- buy one unit at Fr 1,
- sell four units at 50 centimes,
- buy three units at 25 centimes,

would give a profit at all prices provided that the spread between the option at 25 centimes and that at 50 centimes were at most one third of the spread from 50 centimes to Fr 1.[13] We shall see that spreads like this are never seen in practice.

Geometric Representation of Option Transactions[14]

We propose to represent an option purchase geometrically. Let us, for example, take as the origin the price of the forward contract at the time when the option at h has been negotiated. Let E_1 be the relative price of this option or its spread.

Above the exercise price, that is, above the price $(E_1 - h)$ represented by the point A [see Figure 4], the transaction is like a forward negotiated at price E_1. It is therefore represented by the line CBF. Below the price $E_1 - h$, the loss is constant and consequently the transaction is represented by the broken line DCF. The sale of an option is represented in an inverse fashion.

---

[13]The spread here is the difference in the prices in the options, where Bachelier uses *price* to mean the price negotiated in the option, which is the sum of the exercise price and the premium. For this transaction, if the prices of the options at Fr 1, 50 centimes and 25 centimes are, respectively, K_1, K_2 and K_3, the payoff is $(X - (K_1 - 1))_+ - 4(X - (K_2 - 0.5))_+ + 3(X - (K_3 - 0.25))_+ + 0.25$. It is easy to check that this is always positive provided that $(K_2 - K_1) \geqslant 3(K_3 - K_2)$.

[14]The graphical representation of transactions of the exchange is due to Henri Lefèvre, whose books on the stock exchange date from the early 1870s.

True Spreads

Until now we have spoken about quoted spreads, the only ones with which one ordinarily concerns oneself. These are not however the ones that appear in our theory, but instead we consider *true spreads*, that is to say, the spreads between the prices of options and the true prices corresponding to the call date of the options. Since the prices in question are higher than the quoted price (unless the contango is more than 25 centimes, which is rare), it follows that the true spread of an option is less than its quoted spread.

The true spread of an option agreed n days before the call date will be equal to its quoted spread minus the quantity nb. The true spread of an option for the end of the next month will be equal to its quoted spread minus the quantity $[25 + (n - 30)b]$.

Call-of-Mores

In certain markets one agrees transactions that are in some sense intermediate between forward contracts and options, these are the call-of-mores. Let us suppose that the price of a commodity is Fr 30. Instead of buying a single unit for Fr 30 for a given due date, we can buy a double call-of-more for the same due date at Fr 32, for example. This means that for any difference below the price of Fr 32 we only lose one unit while for all differences above, we gain two units.

The double call-of-more can be thought of as giving the buyer the right, but not the obligation, at the liquidation date, to double the number of bonds specified in a forward transaction. Indeed the French term is 'option du double', or, literally, 'option to double'.
The payoff for Bachelier's example is represented geometrically below.

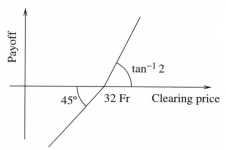

As Bachelier later remarks, one can think of a double call-of-more as the sum of a forward contract and an option.

We could have bought a triple call-of-more at Fr 33, for example, that is to say, for all differences below the price of Fr 33 we lose one unit but for all differences above this price we gain three units. One can imagine multiple-order call-of-mores. The geometric representation of these transactions does not present any difficulty.

One also agrees put-of-mores, necessarily at the same spread as the call-of-mores with the same order of multiplicity.

THE PROBABILITIES IN THE TRANSACTIONS OF THE EXCHANGE

Probabilities in the Transactions of the Exchange

One can consider two types of probability.

1. The probability that one could call *mathematical*, that is, that which one can determine *a priori*. This is what one studies in games of chance.

2. The probability depending on future events, which is, consequently, impossible to predict in a mathematical fashion.

It is this last probability that the speculator tries to predict; he analyses the causes that can influence the rise or the fall of the market and the amplitude of the movements. His inferences are entirely personal, since his counterpart necessarily has the opposite opinion.

It seems that the market, that is to say, the set of speculators, must not believe *at a given instant* in either a rise or a fall, since for each quoted price there are as many buyers as sellers. In reality, the market believes in the rise corresponding to the difference between the coupon payments and the contangoes; the sellers make a small sacrifice that they think of as compensated by this rise. One can ignore this difference by considering the true price corresponding to the liquidation date, but the transactions are regulated by the quoted prices; the seller pays the difference. By considering the true prices one can say:

> At a given instant the market believes neither in a rise nor in a fall of the true price.

But, even if the market believes neither in an increase nor in a decrease of the true price, it can consider movements of a given size to be more or less probable. The object of this study will be the determination of the law of probability that the market accepts at a given time.

THÉORIE DE LA SPÉCULATION

Mathematical Expectation

The product of a final benefit by the corresponding probability of it occurring is called the *mathematical expectation* of the benefit. *The total mathematical expectation* of a player will be the sum of the products of the final benefits and the corresponding probabilities.

Evidently a player will be neither advantaged nor disadvantaged if his total mathematical expectation is zero. One then says that the game is fair.

One knows that bets at the racecourse and all the games practised in casinos are not fair: the casino, or the bookmaker in the case of betting at the races, plays with a positive expectation and the gamblers with a negative expectation. In this sort of game the gamblers cannot choose between the transactions that they make and that taken by their counterpart. Since this is not the case at the Exchange, it may seem curious that these games are not fair; the seller accepts *a priori* a disadvantage if the contangoes are smaller than the coupons. The existence of a second sort of probability explains this apparently paradoxical fact.

The Mathematical Advantage

The mathematical expectation indicates whether a game is advantageous or not. Moreover it teaches us the logical profit or loss of the game, but it does not give a coefficient representing, in some way, the value of the game. This will lead us to introduce a new notion: that of the mathematical advantage.

We call the ratio between the positive expectation and the arithmetic sum of the positive and negative expectations of a player his *mathematical advantage*. Like the probability, the mathematical advantage varies between zero and one, it is equal to $\frac{1}{2}$ if the game is fair.

Principle of Mathematical Expectation

One can liken the spot buyer to a player. Indeed, if the price of a security can go up after the purchase, a fall is equally possible. The causes of this rise or this fall belong to the second class of probabilities.

According to the first type of probability, the price of the security must go up by a value equal to the magnitude of its coupons. Consequently, from the point of view of the first type of probabilities,* the

---

*I consider the simplest case of a security with fixed revenue; if not, the augmentation of the revenue would be a probability of the second class.

mathematical expectation of the spot buyer is positive. If the contango is zero, it is evident that it will be the same as the mathematical expectation of the forward buyer, because this transaction will then be comparable to that of the spot buyer. If the contango on a government bond were 25 centimes, the buyer would be no more advantaged than the seller.

One can therefore say that the mathematical expectations of the buyer and the seller are zero if the contango is 25 centimes. When the contango is less than 25 centimes, which is usually the case, the mathematical expectation of the buyer is positive, that of the seller is negative. It is important to note that this arises solely from the first type of probability.

From what we have already seen, one can always consider the contango as being 25 centimes on replacing the quoted price by the *true* price corresponding to the liquidation date. Thus if one considers true prices, one can say that the mathematical expectations of the buyer and the seller are zero. From the point of view of contangoes, one can consider the call dates of options as identified with the liquidation dates, therefore the mathematical expectations of the buyers and sellers of options are zero. In summary, consideration of true prices allows us to state the fundamental principle:

The mathematical expectation of the speculator is zero[15]

It is important to give an account of the generality of this principle. It signifies that the market, at a given moment, considers as being of zero expectation not only the transactions currently negotiated, but also those that will be based on a subsequent movement of the price. For example, suppose that I buy a government bond with the intention of reselling when it has gained more than 50 centimes. The expectation of this complex transaction is zero, just as if I had the intention of reselling my bond at the liquidation date or any other such time.

The expectation of a transaction cannot be positive or negative unless it is produced by a price movement, *a priori* it is zero.

General Form of the Probability Curve

The probability that the quoted price at a given time will be y is a function of y. One can represent this probability by the ordinate of a curve for which the abscissae will represent different prices.

---

[15]As we pointed out on p. 12, this is tantamount to saying that the true price is a martingale. See also the discussion beginning on p. 91.

It is clear that the price considered by the market to be the most likely is the current true price: if the market judged otherwise it would quote not this price, but another greater or smaller one. For the rest of this study we take the origin of coordinates to be the true price corresponding to the given time. The price can then vary between $-x_0$ and $+\infty$, x_0 being the actual value of the current price. We suppose that it can vary between $-\infty$ and $+\infty$, the probability of a bigger spread than x_0 being considered *a priori* as effectively negligible.[16] Under these conditions, one can assume that the probability of a difference from the true price is independent of the actual value of the price[17] and that the curve of probabilities is symmetric with respect to the true price.

In what follows, we consider only the relative price; the origin of coordinates will always correspond to the current true price.

The Probability Law

The probability law can be determined by the principle of compound probabilities. Let us denote by $p_{x,t}\, dx$ the probability that, at time t, the price is in the elementary interval $x, x + dx$.[18] We seek the probability that the price x will be quoted at time t_1 and the price z will be quoted at time $t_1 + t_2$. By virtue of the principle of composition of probabilities, the required probability will be equal to the product of the probability that the price x will be quoted at the time t_1, that is to say $p_{x,t_1}\, dx$, and the probability that the price z will be quoted at time $t_1 + t_2$ if the price x was quoted at time t_1, that is to say $p_{z-x,t_2}\, dz$. The required probability is therefore

$$p_{x,t_1} p_{z-x,t_2}\, dx\, dz.$$

Since at time t_1 the price can be found in any interval dx between $-\infty$ and $+\infty$, the probability that the price z will be quoted at time $t_1 + t_2$ will be

$$\int_{-\infty}^{+\infty} p_{x,t_1} p_{z-x,t_2}\, dx\, dz.$$

---

[16]This is to combat the problem that otherwise prices could be negative.

[17]In other words, fluctuations about the mean price do not depend on the value of that mean. The geometric Brownian motion model employed in modern mathematical finance makes the same assumption, but about *returns* rather than prices.

[18]We have retained Bachelier's notation for intervals, in modern notation we would write $[x, x + dx)$. He makes no distinction between open and closed intervals and he uses b, a and a, b interchangeably to denote an interval with endpoints a and b.

The probability of price z being quoted at time $t_1 + t_2$ can also be expressed as p_{z,t_1+t_2}. One therefore has

$$p_{z,t_1+t_2}\,\mathrm{d}z = \int_{-\infty}^{+\infty} p_{x,t_1} p_{z-x,t_2}\,\mathrm{d}x\,\mathrm{d}z$$

or

$$p_{z,t_1+t_2} = \int_{-\infty}^{+\infty} p_{x,t_1} p_{z-x,t_2}\,\mathrm{d}x,$$

which is the equation that must be satisfied by the function p.[19] This equation is satisfied, as we shall see, by the function[20]

$$p = A\mathrm{e}^{-B^2 x^2}.$$

Observe now that we must have

$$\int_{-\infty}^{+\infty} p\,\mathrm{d}x = A \int_{-\infty}^{+\infty} \mathrm{e}^{-B^2 x^2}\,\mathrm{d}x = 1.$$

The classical integral that appears in the first term has value $\sqrt{\pi}/B$, therefore $B = A\sqrt{\pi}$ and consequently

$$p = A\mathrm{e}^{-\pi A^2 x^2}.$$

Putting $x = 0$, one obtains $A = p_0$, that is to say A equals the probability of the current quoted price.

It is necessary therefore to establish that the function

$$p = p_0 \mathrm{e}^{-\pi p_0^2 x^2},$$

where p_0 depends on time, satisfies the required equation.

Let p_1 and p_2 be the quantities corresponding to p_0 at times t_1 and t_2; it must be proved that the expression

$$\int_{-\infty}^{+\infty} p_1 \mathrm{e}^{-\pi p_1^2 x^2} \times p_2 \mathrm{e}^{-\pi p_2^2 (z-x)^2}\,\mathrm{d}x$$

---

[19]This is a special case of what is now known as the Chapman–Kolmogorov equation. A more general version can be found in Kolmogorov's 1931 paper 'Über die analytischen Methoden in der Wahrscheinlichkeitsrechnung'. Notice that Bachelier is implicitly assuming that the price process is memoryless (its movements over the interval (t_1, t_2) are independent of those over the interval $(0, t_1)$) and homogeneous in time and space. Kolmogorov credits Bachelier's work as the first of which he is aware that considers stochastic processes in continuous time and refers to this homogeneous setting as 'the Bachelier case'. He remarks, however, 'daß die Bachelierschen Betrachtungen jeder mathematischen Strenge gänzlich entbehren' (that Bachelier's work is wholly without mathematical rigour).

[20]Bachelier is ignoring the issue of uniqueness of solutions to the equation here. As Kolmogorov showed, under his (implicit) assumptions, Bachelier's conclusions in this section are correct, but his proof is not rigorous.

can be written in the form $Ae^{-\pi B^2 z^2}$ with A and B depending only on time. Noting that z is a constant, this integral becomes

$$p_1 p_2 e^{-\pi p_2^2 z^2} \int_{-\infty}^{+\infty} e^{-\pi (p_1^2 + p_2^2)x^2 + 2\pi p_2^2 zx}\, dx$$

or

$$p_1 p_2 \exp\left[-\pi p_2^2 z^2 + \frac{\pi p_2^4 z^2}{p_1^2 + p_2^2} \right]$$
$$\times \int_{-\infty}^{+\infty} \exp\left[-\pi \left(x\sqrt{p_1^2 + p_2^2} - \frac{p_2^2 z}{\sqrt{p_1^2 + p_2^2}} \right)^2 \right] dx.$$

Let us set

$$x\sqrt{p_1^2 + p_2^2} - \frac{p_2^2 z}{\sqrt{p_1^2 + p_2^2}} = u.$$

We will then have

$$\frac{p_1 p_2}{\sqrt{p_1^2 + p_2^2}} \exp\left[-\pi p_2^2 z^2 + \frac{\pi p_2^4 z^2}{p_1^2 + p_2^2} \right] \int_{-\infty}^{+\infty} e^{-\pi u^2}\, du.$$

The integral having value 1, we finally obtain

$$\frac{p_1 p_2}{\sqrt{p_1^2 + p_2^2}} \exp\left[-\pi \frac{p_1^2 p_2^2}{p_1^2 + p_2^2} \right].$$

This expression has the desired form and so we must conclude from this that the probability can be expressed by the formula

$$p = p_0 e^{-\pi p_0^2 x^2},$$

in which p_0 depends on time.

One sees that the probability is regulated by the Gauss law, already famous in the calculus of probabilities.

Probability as a Function of Time

The last but one formula shows us that the parameters $p_0 = f(t)$ satisfy the functional relationship

$$f^2(t_1 + t_2) = \frac{f^2(t_1) f^2(t_2)}{f^2(t_1) + f^2(t_2)}.$$

31

Let us differentiate with respect to t_1, then with respect to t_2. The first term[21] having the same form in both cases, we obtain

$$\frac{f'(t_1)}{f^3(t_1)} = \frac{f'(t_2)}{f^3(t_2)}.$$

Since this relation holds for any t_1 and t_2, the common value of the two ratios is constant and we have

$$f'(t) = Cf^3(t),$$

from which

$$f(t) = p_0 = \frac{H}{\sqrt{t}},$$

H denoting a constant.

We therefore have as an expression for the probability[22]

$$p = \frac{H}{\sqrt{t}} \exp\left[- \frac{\pi H^2 x^2}{t} \right].$$

Mathematical Expectation

The expectation corresponding to the price x has value

$$\frac{Hx}{\sqrt{t}} \exp\left[- \frac{\pi H^2 x^2}{t} \right].$$

The total positive expectation is therefore

$$\int_0^\infty \frac{Hx}{\sqrt{t}} \exp\left[- \frac{\pi H^2 x^2}{t} \right] dx = \frac{\sqrt{t}}{2\pi H}.$$

In our study we take as constant the positive mathematical expectation k corresponding to $t = 1$. We therefore have

$$k = \frac{1}{2\pi H} \quad \text{or} \quad H = \frac{1}{2\pi k}.$$

The definitive expression for the probability is therefore

$$p = \frac{1}{2\pi k\sqrt{t}} \exp\left[- \frac{x^2}{4\pi k^2 t} \right].$$

---

[21] That is, the derivative of the left-hand side.

[22] This is the probability density function of a normally distributed random variable with variance $t/2\pi H^2$. The price process is what we would now call a Brownian motion with variance parameter $1/2\pi H^2$. In 1905, Einstein, unaware of Bachelier's work, derived the same expression for the transition density of Brownian motion in his paper 'On the motion of particles suspended in fluids at rest implied by the molecular–kinetic theory of heat'.

The positive mathematical expectation

$$\int_0^\infty px\,\mathrm{d}x = k\sqrt{t}$$

is proportional to the square root of time.

New Determination of the Probability Law[23]

The expression for the function p can be obtained via a different route from that which we have followed.

I suppose that two complementary events A and B have probabilities p and $q = 1 - p$, respectively. The probability that in a string of m such events, α are equal to A and $m - \alpha$ are equal to B is given by the expression

$$\frac{m!}{\alpha!(m-\alpha)!}p^\alpha q^{m-\alpha}.$$

This is one of the terms in the expansion of $(p+q)^m$. The largest of these probabilities occurs when

$$\alpha = mp \quad \text{and} \quad (m-\alpha) = mq.$$

Let us consider the term for which the exponent of p is $mp + h$. The corresponding probability is

$$\frac{m!}{(mp+h)!(mq-h)!}p^{mp+h}q^{mq-h}.$$

The quantity h is called the *spread*.

Let us look for the mathematical expectation of a player who will receive a sum equal to the spread if this spread is positive. We have just seen that the probability of a spread h is the term in the expansion of $(p+q)^m$ in which the exponent of p is $mp+h$ and that of q is $mq-h$. To obtain the mathematical expectation corresponding to this term, it will be necessary to multiply this probability by h. Now

$$h = q(mp+h) - p(mq-h),$$

---

[23]This section contains many errors noted in the footnotes. However, the basic idea is correct. Bachelier is thinking of Brownian motion as an 'infinitesimal limit' of random walks and, since he is concerned only with the probability law at a fixed time, he is essentially reproducing the normal approximation to the binomial distribution due to de Moivre (presented in Latin in a printed note in 1733 and translated into English in 1738 in the second edition of his book *The Doctrine of the Chances*) or a special form of Laplace's (1810) version of the Central Limit Theorem.

and $mp + h$ and $mq - h$ are the exponents of p and of q in the corresponding term of $(p + q)^m$. To multiply a term

$$q^\mu p^\nu$$

by

$$\nu q - \mu p = pq\left(\frac{\nu}{p} - \frac{\mu}{q}\right)$$

is to take the derivative with respect to p, subtract the derivative with respect to q and multiply the difference by pq. To obtain the total positive mathematical expectation, we must therefore take the terms in the expansion of $(p + q)^m$ for which h is positive, that is to say

$$p^m + mp^{m-1}q + \frac{m(m-1)}{1.2}p^{m-2}q^2 + \cdots \frac{m!}{mp!mq!}p^{mp}q^{mq},$$

subtract the derivative with respect to q from the derivative with respect to p, and then multiply the result by pq.

The derivative of the second term with respect to q is equal to the derivative of the first with respect to p, the derivative of the third with respect to q is the derivative of the second with respect to p, and so on. The terms therefore cancel one another two by two and all that remains is the derivative of the last term with respect to p,[24]

$$\frac{m!}{mp!mq!}p^{mp}q^{mq}mpq.$$

The average magnitude of the spread h will be equal to twice this quantity.

Provided the number m is sufficiently large, one can simplify the preceding expressions by making use of the asymptotic formula of Stirling,

$$n! = e^{-n}n^n\sqrt{2\pi n}.$$

In this way, one obtains the value

$$\frac{\sqrt{mpq}}{\sqrt{2\pi}}$$

for the positive mathematical expectation. The probability that the spread h will be contained between h and $h + dh$ is given by the expression[25]

$$\frac{dh}{\sqrt{2\pi mpq}}\exp\left[-\frac{h^2}{2mpq}\right].$$

---

[24] The formula now given is that derivative multiplied by pq; that is, the total positive mathematical expectation.

[25] Since h can only take integer values, the notion dh does not make much sense. This is the root of many of the errors that follow.

We can apply the preceding theory to our study. One can suppose that time is divided into very small intervals Δt, in such a way that $t = m\Delta t$. During the time Δt the price will probably vary very little. Let us form the sum of the products of the spreads that can exist at the time Δt by the corresponding probabilities; that is to say, $\int_0^\infty px\,dx$, p being the probability of the spread x. This integral must be finite because, as a result of the smallness assumed for Δt, large spreads have a vanishing probability. Moreover, this integral expresses a mathematical expectation, which cannot but be finite if it corresponds to a very small interval of time.

Let us denote by Δx twice the value of the integral above. Δx will be the average of the spreads, or the mean spread, during the time Δt.[26]

If the number m of trials is very big and the probability remains the same for each trial, we can suppose that the price varies for each of the trials Δt by the mean spread Δx. The rise Δx has probability $\frac{1}{2}$ as has the fall $-\Delta x$. The preceding formula therefore gives, on putting $p = q = \frac{1}{2}$, the probability that, at time t, the price will be between x and $x + dx$. This will be[27]

$$\frac{2\,dx\sqrt{\Delta t}}{\sqrt{2\pi}\sqrt{t}}\exp\left[-\frac{2x^2\Delta t}{t}\right],$$

or, putting $H = (2/\sqrt{2\pi})\sqrt{\Delta t}$,[28]

$$\frac{H\,dx}{\sqrt{t}}\exp\left[-\frac{\pi H^2 x^2}{t}\right].$$

The positive mathematical expectation has the expression

$$\frac{\sqrt{t}}{2\sqrt{2\pi}\sqrt{\Delta t}} = \frac{\sqrt{t}}{2\pi H}.$$

If we take as constant the positive mathematical expectation k corresponding to $t = 1$, we find, as before,

$$p = \frac{1}{2\pi k\sqrt{t}}\exp\left[-\frac{x^2}{4\pi k^2 t}\right].$$

---

[26]Here 'mean spread' really means the average of the absolute value of the spread.

[27]Bachelier has not substituted the right value for h here: since

$$(mp + h)\Delta x - (mq - h)\Delta x = m(p - q)\Delta x + 2h\Delta x,$$

a spread of h results in a deviation of $2h\Delta x$ from the mean price. For a spread of x in the price we therefore need $h = x/(2\Delta x)$. This gives the probability that the price is in $[x, x + dx)$ as

$$\frac{dx\sqrt{\Delta t}}{\Delta x\sqrt{2\pi}\sqrt{t}}\exp\left[-\frac{x^2\Delta t}{2(\Delta x)^2 t}\right].$$

[28]Set $H = \sqrt{\Delta t}/\sqrt{2\pi}\Delta x$ to recover this from the correct expression.

The preceding formulae give $\Delta t = 1/8\pi k^2$.[29] The average spread during this time interval is[30]

$$\Delta x = \frac{\sqrt{2}}{2\sqrt{\pi}}.$$

If we put $x = n\Delta x$, the probability has expression[31]

$$p = \frac{\sqrt{2}}{\sqrt{\pi}\sqrt{m}} \exp\left[-\frac{n^2}{\pi m}\right].$$

Curve of Probabilities

The function

$$p = p_0 e^{-\pi p_0^2 x^2}$$

can be represented by a curve for which the ordinate is maximized at the origin and which presents two points of inflexion for

$$x = \pm\frac{1}{p_0\sqrt{2\pi}} = \pm\sqrt{2\pi}k\sqrt{t}.$$

These same values of x are the x-coordinates of the maxima and minima of the curve of mathematical expectation, for which the equation is[32]

$$y = \pm px.$$

The probability of the price x is a function of t. It increases until a certain time and then it decreases. The derivative $dp/dt = 0$ when $t = x^2/2\pi k^2$. The probability of the price x is therefore maximized when this price corresponds to the point of inflexion of the probability curve.

---

[29]This reads $\Delta t = (\Delta x)^2/2\pi k^2$ if the correct value of H is substituted.

[30]We can see that something has gone wrong as the quantities Δt and Δx given by these expressions are not 'very small' as we have assumed.

[31]This formula is not consistent with that given earlier for the probability of a 'spread' in $h, h + dh$. However, if we set $x = 2n\Delta x$ and $t = m\Delta t$ then we do recover

$$p = \frac{2}{\sqrt{2\pi m}} \exp\left[-\frac{2n^2}{m}\right]$$

as before (where n plays the role of h).

[32]Again Bachelier is concerned with the absolute value of the spread. The two equations for the mathematical expectation are because he considers positive and negative values of x separately.

Probability in a Given Interval[33]

The integral

$$\frac{1}{2\pi k \sqrt{t}} \int_0^x \exp\left[-\frac{x^2}{4\pi k^2 t}\right] dx = \frac{c}{\sqrt{\pi}} \int_0^x e^{-c^2 x^2}\, dx$$

is not expressible in finite terms, but one can give its series expansion:

$$\frac{1}{\sqrt{\pi}}\left[cx - \frac{\frac{1}{3}(cx)^3}{1} + \frac{\frac{1}{5}(cx)^5}{1.2} - \frac{\frac{1}{7}(cx)^7}{1.2.3} + \cdots\right].$$

This series converges quite slowly for very large values of cx. For this case Laplace has expressed the definite integral in the form of a continued fraction that is very easy to calculate:

$$\frac{1}{2} - \frac{e^{-c^2 x^2}}{2cx\sqrt{\pi}} \cfrac{1}{1 + \cfrac{\alpha}{1 + \cfrac{2\alpha}{1 + \cfrac{3\alpha}{1 + \cdots}}}},$$

in which $\alpha = 1/2c^2 x^2$. The successive reductions are

$$\frac{1}{1+\alpha}, \quad \frac{1+2\alpha}{1+3\alpha}, \quad \frac{1+5\alpha}{1+6\alpha+3\alpha^2}, \quad \frac{1+9\alpha+8\alpha^2}{1+10\alpha+15\alpha^2}.$$

There is another procedure that allows the calculation of the integral above when x is a large number. One has

$$\int_x^\infty e^{-x^2}\, dx = \int_x^\infty \frac{1}{2x} e^{-x^2} 2x\, dx.$$

On integrating by parts, one then obtains

$$\int_x^\infty e^{-x^2}\, dx = \frac{e^{-x^2}}{2x} - \int_x^\infty e^{-x^2} \frac{dx}{2x^2}$$

$$= \frac{e^{-x^2}}{2x} - \frac{e^{-x^2}}{4x^3} + \int_x^\infty e^{-x^2} \frac{1.3}{4x^4}\, dx$$

$$= \frac{e^{-x^2}}{2x} - \frac{e^{-x^2}}{4x^3} + \frac{e^{-x^2}1.3}{8x^5} - \int_x^\infty e^{-x^2} \frac{1.3.5}{8x^6}\, dx.$$

---

[33]In this section, Bachelier provides a number of approaches to estimating the error function, although, as he remarks, published tables were available. Here, and elsewhere, we have preserved his somewhat informal notation and allowed x to appear in both the integrand and the upper limit of the integral.

The general term in the series is

$$\frac{1 \cdot 3 \cdot 5 \cdots \cdots (2n-1)}{2^{2n-1} x^{2n+1}} e^{-x^2}.$$

The ratio of one term to the previous term is greater than one if $2n+1 > 4x^2$. The series therefore diverges after a certain term. One can obtain a limit supremum of the integral that serves as the remainder. Indeed one has

$$\frac{1 \cdot 3 \cdot 5 \cdots \cdots (2n-1)}{2^{2n-1}} \int_x^\infty \frac{e^{-x^2}}{x^{2n+2}} \, dx$$

$$< \frac{1 \cdot 3 \cdot 5 \cdots \cdots (2n-1)}{2^{2n-1}} e^{-x^2} \int_x^\infty \frac{1}{x^{2n+2}} \, dx$$

$$= \frac{1 \cdot 3 \cdot 5 \cdots \cdots (2n-1)}{2^{2n-1} x^{2n+1}} e^{-x^2}.$$

But this last quantity is the term that preceded the integral. The complementary term is therefore always smaller than that which precedes it.

There are published tables giving the values of the integral

$$\Theta(y) = \frac{2}{\sqrt{\pi}} \int_0^y e^{-y^2} \, dy.$$

One will evidently have

$$\int_0^x p \, dx = \tfrac{1}{2} \Theta \left(\frac{x}{2k\sqrt{\pi}\sqrt{t}} \right).$$

The probability

$$\mathcal{P} = \int_x^\infty p \, dx = \frac{1}{2} - \frac{1}{2} \frac{2}{\sqrt{\pi}} \int_0^{x/(2\sqrt{\pi}k\sqrt{t})} e^{-\lambda^2} \, d\lambda$$

that the price will be greater than or equal to x at time t increases constantly with time. If t were infinite, it would be equal to $\frac{1}{2}$, an obvious result.

The probability

$$\int_{x_1}^{x_2} p \, dx = \frac{1}{\sqrt{\pi}} \int_{x_1/(2\sqrt{\pi}k\sqrt{t})}^{x_2/(2\sqrt{\pi}k\sqrt{t})} e^{-\lambda^2} \, d\lambda$$

that the price is contained in the interval x_2, x_1 at time t is zero for $t = 0$ and for $t = \infty$. It is maximized when

$$t = \frac{1}{4\pi k^2} \frac{x_2^2 - x_1^2}{\log(x_2/x_1)}.$$

If we suppose that the interval x_2, x_1 is very small, we recover the time when the probability is maximized:

$$t = \frac{x^2}{2\pi k^2}.$$

Probable Spread

This is what we call the interval $\pm\alpha$ such that at the end of time t the price has as much chance of staying inside the interval as of being outside.

The quantity α is determined by the equation

$$\int_0^\alpha p\,dx = \tfrac{1}{4}$$

or

$$\Theta\left(\frac{\alpha}{2k\sqrt{\pi}\sqrt{t}}\right) = \tfrac{1}{2},$$

that is to say,

$$\alpha = 2 \times 0.4769k\sqrt{\pi}\sqrt{t} = 1.688k\sqrt{t};$$

this interval is proportional to the square root of time.

More generally, let us consider the interval $\pm\beta$ such that the probability that, at time t, the price will be contained in this interval will be equal to u. We will have

$$\int_0^\beta p\,dx = \tfrac{1}{2}u$$

or

$$\Theta\left(\frac{\beta}{2k\sqrt{\pi}\sqrt{t}}\right) = u.$$

We see that this interval is proportional to the square root of time.

Radiation of the Probability[34]

I am going to seek directly the expression for the probability $\mathcal{P}$ that the price will be greater than or equal to x at time t. We have seen above that by dividing time into very small intervals Δt, one can consider, during an interval Δt, the price as varying by the fixed and very small quantity Δx. I suppose that, at time t, the prices $x_{n-2}, x_{n-1}, x_n, x_{n+1}, x_{n+2}, \ldots$, with differences Δx between them, have respective probabilities $p_{n-2}, p_{n-1}, p_n, p_{n+1}, p_{n+2}, \ldots$. From the knowledge of the distribution of probabilities at time t, one easily deduces the distribution of probabilities at time $t + \Delta t$. Let us suppose, for example, that the price x_n is quoted at time t. At the time $t + \Delta t$ the price

---

[34]In this section Bachelier uses a discretization scheme to make the connection with Fourier's theory of radiation (flux of heat) and hence the heat equation. He thinks of 'probability flux' in place of heat flux.

x_{n+1} or x_{n-1} will be quoted. The probability p_n that the price x_n will be quoted at time t decomposes into two probabilities at time $t + \Delta t$ and from this fact the price x_{n-1} has probability $\frac{1}{2}p_n$ and the price x_{n+1} also has probability $\frac{1}{2}p_n$.

If the price x_{n-1} is quoted at time $t + \Delta t$, the price x_{n-2} or x_n must have been quoted at time t. The probability of the price being x_{n-1} at time $t + \Delta t$ is therefore $\frac{1}{2}(p_{n-2} + p_n)$, that of the price being x_n at the same time is $\frac{1}{2}(p_{n-1} + p_{n+1})$, that of the price being x_{n+1} is $\frac{1}{2}(p_n + p_{n+2})$, etc.

During the time Δt, the price x_n has, in some way, emitted towards the price x_{n+1} the probability $\frac{1}{2}p_n$; the price x_{n+1} has emitted towards the price x_n the probability $\frac{1}{2}p_{n+1}$. If p_n is bigger than p_{n+1}, the exchange in probabilities is $\frac{1}{2}(p_n - p_{n+1})$ from x_n to x_{n+1}. One can therefore say:

> *During the time element Δt, each price x radiates a quantity of probability proportional to the difference in their probabilities towards the neighbouring price.*

I say proportional, because one must take account of the ratio of Δx to Δt.

The preceding law can, by analogy with certain physical theories, be called the *law of radiation* or the diffusion of the probability.

I now consider the probability $\mathcal{P}$ that at time t the price x is in the interval x, ∞ and I evaluate the increase in this probability during the time Δt. Let p be the probability of price x at time t, $p = -\mathrm{d}\mathcal{P}/\mathrm{d}x$. Let us evaluate the probability which passes, in some sense, through the price x during the time interval Δt. From what has just been said this is

$$\frac{1}{c^2}\left(p - \frac{\mathrm{d}p}{\mathrm{d}x} - p\right)\Delta t = -\frac{1}{c^2}\frac{\mathrm{d}p}{\mathrm{d}x}\Delta t = \frac{1}{c^2}\frac{\mathrm{d}^2\mathcal{P}}{\mathrm{d}x^2}\Delta t,$$

where c denotes a constant.[35] This increase in probability can also be expressed as $(\mathrm{d}\mathcal{P}/\mathrm{d}t)\Delta t$. One therefore has

$$c^2\frac{\partial\mathcal{P}}{\partial t} - \frac{\partial^2\mathcal{P}}{\partial x^2} = 0.$$

This is an equation of Fourier.[36]

---

[35] This is most easily understood by thinking of x as lying between x_n and x_{n+1} in the argument above. The 'flux' passing through x is then proportional to $p_n - p_{n+1}$ times the length of the time interval. If p is the probability at x_n, then the probability at x_{n+1} is (to first order) $p + \Delta x(\mathrm{d}p/\mathrm{d}x)$, so the flux is proportional to $-(\mathrm{d}p/\mathrm{d}x)\Delta t$.

[36] This is what we now call the heat equation. Fourier's law says that heat flow is proportional to the temperature gradient. Bachelier has shown that probability flow is proportional to the probability gradient.

The preceding theory assumes that the variations in the price are discontinuous. One can arrive at the equation of Fourier without making this hypothesis by observing that in a very small interval of time Δt the price varies in a continuous way, but by a very small quantity, which is less than ϵ, say. We denote by ω the probability corresponding to p and relative to Δt. From our hypothesis, the price cannot vary except within the limits $\pm\epsilon$ in the time Δt and consequently

$$\int_{-\epsilon}^{+\epsilon} \omega \, dx = 1.$$

If m is positive and smaller than ϵ, the price can be $x - m$ at time t. The probability of this event is p_{x-m}. The probability that the price x will be exceeded at the time $t + \Delta t$, having been equal to $x - m$ at time t, will, by virtue of the principle of composition of probabilities, have value[37]

$$p_{x-m} \int_{\epsilon-m}^{\epsilon} \omega \, dx.$$

The price could be $x + m$ at time t; the probability of this event is p_{x+m}. The probability that the price will be less than x at time $t + \Delta t$, having been equal to $x+m$ at time t, has value, by virtue of the principle invoked before,[38]

$$p_{x+m} \int_{\epsilon-m}^{\epsilon} \omega \, dx.$$

The increase in the probability $\mathcal{P}$, in the time interval Δt, will be equal to the sum of the expressions

$$(p_{x-m} - p_{x+m}) \int_{\epsilon-m}^{\epsilon} \omega \, dx$$

for all values of m from zero to ϵ.

Let us expand the expressions for p_{x-m} and p_{x+m}. Ignoring terms that contain m^2 we will have

$$p_{x-m} = p_x - m\frac{dp_x}{dx},$$
$$p_{x+m} = p_x + m\frac{dp_x}{dx}.$$

---

[37]The lower limit of integration should be m in the each of the next three integrals.

[38]Bachelier is once again assuming symmetry of the distribution of the spread around the mean here.

The expression above then becomes[39]

$$-\frac{\mathrm{d}p}{\mathrm{d}x} \int_{\epsilon-m}^{\epsilon} 2m\omega\,\mathrm{d}x.$$

The required increase therefore has value

$$-\frac{\mathrm{d}p}{\mathrm{d}x} \int_{0}^{\epsilon} \int_{\epsilon-m}^{\epsilon} 2m\omega\,\mathrm{d}x\,\mathrm{d}m.$$

The integral does not depend on x, on t or on p, it is a constant. The increase in the probability $\mathcal{P}$ is therefore[40]

$$\frac{1}{c^2}\frac{\mathrm{d}p}{\mathrm{d}x}.$$

Fourier's equation has integral

$$\mathcal{P} = \int_{0}^{\infty} f\left(t - \frac{c^2 x^2}{2\alpha^2}\right) e^{-\alpha^2/2}\,\mathrm{d}\alpha.$$

The arbitrary function f is determined by the following considerations: for any positive value of t one must have $\mathcal{P} = \frac{1}{2}$ if $x = 0$ and $\mathcal{P} = 0$ if t is negative. Putting $x = 0$ in the integral above, we have

$$\mathcal{P} = f(t) \int_{0}^{\infty} e^{-\alpha^2/2}\,\mathrm{d}\alpha = \frac{\sqrt{\pi}}{\sqrt{2}} f(t),$$

that is to say,

$$f(t) = \begin{cases} \dfrac{1}{\sqrt{2}\sqrt{\pi}} & \text{for } t > 0, \\ 0 & \text{for } t < 0. \end{cases}$$

This last equality shows us that the integral $\mathcal{P}$ has zero elements as long as $t - (c^2 x^2/2\alpha^2)$ is smaller than zero, that is as long as α is smaller than $cx/\sqrt{2}\sqrt{t}$. One must therefore take the quantity $cx/\sqrt{2}\sqrt{t}$ for the lower limit in the integral $\mathcal{P}$, and one then has

$$\mathcal{P} = \frac{1}{\sqrt{2}\sqrt{\pi}} \int_{cx/(\sqrt{2}\sqrt{t})}^{\infty} e^{-\alpha^2/2}\,\mathrm{d}\alpha = \frac{1}{\sqrt{\pi}} \int_{cx/(2\sqrt{t})}^{\infty} e^{-\lambda^2}\,\mathrm{d}\lambda,$$

or, replacing $\int_{cx/(2\sqrt{t})}^{\infty}$ by $\int_{0}^{\infty} - \int_{0}^{cx/(2\sqrt{t})}$,

$$\mathcal{P} = \frac{1}{2} - \frac{1}{2}\frac{2}{\sqrt{\pi}} \int_{0}^{cx/(2\sqrt{t})} e^{-\lambda^2}\,\mathrm{d}\lambda,$$

the formula we found before.

---

[39] Again the lower limit of integration should be m both here and in the inner integral of the following equation.

[40] In fact it is minus the expression that Bachelier now gives.

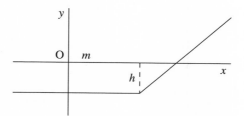

Figure 5.

Law of the Spreads of Options

In order to find the law that regulates the relationship between the premiums of options and their spreads, we apply the principle of mathematical expectation to the buyer of the option:

The mathematical expectation of the buyer of the option is zero.

We take as the origin the true price of the forward (Figure 5). Let p be the probability of price $\pm x$, that is, in the current case, the probability that the price at the call date of the option is $\pm x$. Let $m + h$ be the true spread of the option at h.[41] We spell out the condition that the total mathematical expectation is zero.

We evaluate this expectation

1. for prices between $-\infty$ and m,
2. for prices between m and $m + h$,
3. for prices between $m + h$ and ∞.

1. For all prices between $-\infty$ and m, the option is abandoned, that is to say the buyer accepts a loss of h. His mathematical expectation for a price in the given interval is therefore $-ph$ and, for the whole interval,

$$-h \int_{-\infty}^{m} p \, dx.$$

2. For a price x between m and $m + h$, the loss of the buyer will be $m + h - x$. The corresponding mathematical expectation will be $p(m+h-x)$ and, for the whole interval,

$$-\int_{m}^{m+h} p(m + h - x) \, dx.$$

3. For a price x between $m + h$ and ∞, the profit to the buyer will be $x - m - h$. The corresponding mathematical expectation will be

---

[41] Recall the definition from p. 25.

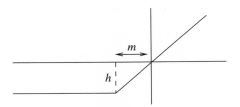

Figure 6.

$p(x - m - h)$ and, for the whole interval,

$$\int_{m+h}^{\infty} p(x - m - h)\, dx.$$

The principle of total expectation will therefore give

$$\int_{m+h}^{\infty} p(x - m - h)\, dx - \int_{m}^{m+h} p(m + h - x)\, dx - h \int_{-\infty}^{m} p\, dx = 0$$

or, simplifying,

$$h + m \int_{m}^{\infty} p\, dx = \int_{m}^{\infty} px\, dx.$$

This is the equation in definite integrals that establishes a relationship between the probabilities, the spreads of options and their premiums.

In the case where the exercise price of the option falls to the side of negative x, as illustrated in Figure 6, m will be negative and one arrives at the relation[42]

$$\frac{2h + m}{2} + m \int_{0}^{-m} p\, dx = \int_{-m}^{\infty} px\, dx.$$

As a result of the symmetry of the probabilities the function p must be even and so the two equations above are really a single equation.

Differentiating, one obtains the differential equation governing the spreads of the option,[43]

$$\frac{d^2 h}{dm^2} = p_m,$$

p_m being the expression for the probability evaluated with $x = m$.

---

[42] It is not clear why Bachelier feels the need to treat negative m separately. In fact he has made a mistake and the lower limit of the integral on the right-hand side should be m rather than $-m$.

[43] The first equation gives $h = \int_{m}^{\infty} p(x - m)\, dx$, so

$$\frac{dh}{dm} = -\int_{m}^{\infty} p\, dx$$

and differentiating again the result follows.

Simple Option

The simplest case of the equations above is when $m = 0$, that is to say that when the premium of the option is equal to its spread.[44] This sort of option, the only one that is negotiated in speculation on commodities, is called a *simple option*. On putting $m = 0$ and denoting the premium of the simple option by a the equations above become

$$a = \int_0^\infty px\,dx = \int_0^\infty \frac{x}{2\pi k\sqrt{t}} \exp\left[-\frac{x^2}{4\pi k^2 t}\right] dx = k\sqrt{t}.$$

The equality $a = \int_0^\infty px\,dx$ shows that the premium of the simple option is equal to the positive expectation of a forward buyer. This fact is evident, since the buyer of the option pays the sum a to the seller to enjoy the advantages of the forward buyer, that is, to have his positive expectation without incurring his risks.

From the formula

$$a = \int^\infty px\,dx = k\sqrt{t},$$

we deduce the following principle, one of the most important of our study:

The value of the simple option must be proportional to the square root of time.

We have already seen that the probable spread is given by the formula

$$\alpha = 1.688k\sqrt{t} = 1.688a.$$

The probable spread is therefore found by multiplying the average of the option by the constant 1.688. It is therefore very easy to calculate when it arises from speculation on commodities as, in this case, the quantity a is known.

The following formula gives the expression for the probability of the price x as a function of a:

$$p = \frac{1}{2\pi a} \exp\left[-\frac{x^2}{4\pi a^2}\right].$$

The probability of being in a given interval has the integral expression

$$\frac{1}{2\pi a} \int_0^u \exp\left[-\frac{x^2}{4\pi a^2}\right] dx$$

---

[44]That is, the exercise price equals the forward price.

or

$$\frac{1}{2\pi a}\left(u - \frac{u^3}{12\pi a^2} + \frac{u^5}{160\pi^2 a^4} - \frac{u^7}{2678\pi^3 a^6} + \cdots\right).$$

If u, instead of being a given number, is a parameter of the form $u = ba$, then this probability is independent of a and consequently of time. For example, if $u = a$,

$$\int_0^a p\,\mathrm{d}x = \frac{1}{2\pi} - \frac{1}{24\pi^2} + \frac{1}{320\pi^3} - \cdots = 0.155.$$

The integral $\int_a^\infty p\,\mathrm{d}x$ represents the probability of profit for the buyer of the simple option. Now

$$\int_a^\infty p\,\mathrm{d}x = \tfrac{1}{2} - \int_0^a p\,\mathrm{d}x = 0.345.$$

Therefore,

The probability of profit for the buyer of the simple option is independent of the time of maturity. It has value

0.345.

The positive expectation of the simple option is

$$\int_a^\infty p(x - a)\,\mathrm{d}x = 0.58a.$$

Double Options[45]

The *put and call option* or *double option* is formed by the simultaneous purchase of an upside option and of a downside option (simple options). It is easy to see that the seller of a double option is in profit in the interval $-2a, +2a$. His probability of profit is therefore

$$2\int_0^{2a} p\,\mathrm{d}x = \frac{2}{\pi} - \frac{2}{3\pi^2} + \frac{2}{10\pi^3} - \cdots = 0.56.$$

The probability of profit for the buyer of the double option is 0.44.

The positive expectation of the double option is

$$2\int_{2a}^\infty p(x - 2a)\,\mathrm{d}x = 0.55a.$$

---

[45] This transaction, called a *stellage* or *double prime* in French, is now most commonly called a *straddle*.

Coefficient of Instability

The coefficient k that we introduced previously is the *coefficient of instability*[46] or of the excitability of the security; it measures its static state. Its strength indicates a state of anxiety; its weakness, by contrast, is the indicator of a calm state.

This coefficient is given directly in speculation on commodities by the formula

$$a = k\sqrt{t},$$

but in speculation on securities, as we shall see, one cannot calculate it except by approximation.

Series Expansion for the Spread of an Option

The equation in terms of definite integrals for the spread of an option is not expressible in finite terms when the quantity m, the difference between the spread of the option and its premium h, is non-zero. The equation leads to the series

$$h - a + \frac{m}{2} - \frac{m^2}{4\pi a} + \frac{m^4}{96\pi^2 a^3} - \frac{m^6}{1920\pi^3 a^5} + \cdots = 0.$$

This relation, in which the quantity a denotes the premium of the simple option, allows one to calculate the value of a when one knows that of m or vice versa.

Approximate Law of the Spreads of the Option

The preceding series can be written

$$h = a - f(m).$$

Let us consider the product of the premium, h, of the option and its spread $(m + h)$:

$$h(m + h) = [a - f(m)][m + a - f(m)];$$

differentiating in m we have[47]

$$\frac{d}{dm}[h(m + h)] = f'(m)[m + a - f(m)] + [a - f(m)][1 - f'(m)].$$

---

[46] It plays the role of the *volatility* in modern mathematical finance.

[47] In fact the first term on the right-hand side of the expression that Bachelier now gives should be multiplied by -1.

47

If we put $m = 0$, from which $f(m) = 0$, $f'(m) = \frac{1}{2}$, this derivative is zero. We must conclude from this that:

> *The product of the premium of an option and its spread is maximized when the two factors of the product are equal. This is the case for a simple option.*

In the neighbourhood of its maximum, the product in question must vary little. It is this that often allows the approximate determination of a by the formula

$$h(m + h) = a^2,$$

which gives too small a value for a.

By considering only the first three terms in the series one obtains

$$h(h + m) = a^2 - \tfrac{1}{4}m^2,$$

which gives too large a value for a.

In most cases, by taking the first four terms of the series one would obtain a very adequate approximation. In this way one has

$$a = \frac{\pi(2h + m) \pm \sqrt{\pi^2(2h + m)^2 - 4\pi m^2}}{4\pi}.$$

With this same approximation one has for the value of m as a function of a

$$m = \pi a \pm \sqrt{\pi^2 a^2 - 4\pi a(a - h)}.$$

Let us assume for a moment the simpler formula

$$h(m + h) = a^2 = k^2 t.$$

In speculation on securities, upside options have a constant premium h, the spread $m + h$ is therefore proportional to time.

> *In speculation on securities, the spread of upside options is directly proportional to the duration of their maturity and the square of the instability.*

Downside options on securities (that is to say, the sale of a forward contract against the purchase of an option) have a constant spread h and a variable premium $m + h$.

> *In speculation on securities, the premium of downside options is directly proportional to their maturity and to the square of the instability.*

The two laws above are only approximate.

Call-of-Mores

Let us apply the principle of mathematical expectation to the purchase of a call-of-more of order n agreed at a spread r. The call-of-more of order n can be thought of as composed of two transactions:

1. the purchase of a forward contract for one unit at price r,
2. the purchase of a forward contract for $(n-1)$ units at price r, this purchase only being considered in the interval r, ∞.

The first transaction has mathematical expectation $-r$,[48] the second has expectation

$$(n-1) \int_r^\infty p(x-r)\,dx.$$

One must therefore have

$$r = (n-1) \int_r^\infty p(x-r)\,dx$$

or, replacing p by its value,

$$p = \frac{1}{2\pi a} \exp\left[-\frac{x^2}{4\pi a^2} \right],$$

and developing as a series,[49]

$$2\pi a^2 - \pi a \frac{n+1}{n-1} r + \frac{r^2}{2} - \frac{r^4}{48\pi a^2} + \cdots = 0.$$

Conserving only the first three terms we obtain

$$r = a\left[\frac{n+1}{n-1}\pi - \sqrt{\left(\frac{n+1}{n-1}\pi\right)^2 - 4\pi} \right].$$

If $n = 2$,

$$r = 0.68a.$$

The spread of the double call-of-more must be about two thirds of the value of the simple option.

If $n = 3$,

$$r = 1.096a.$$

The spread of the call-of-more of order three must be about one tenth greater than the value of the simple option.

---

[48]Recall that the origin of coordinates is the true price.

[49]To see this, write $\int_r^\infty p(x-r)\,dx$ as $(\int_0^\infty - \int_0^r)p(x-r)\,dx$, substitute $\int_0^\infty px\,dx = a$, $\int_0^\infty pr\,dx = \frac{1}{2}r$, and for the integral over $[0, r]$ develop $e^{-x^2/4\pi a^2}$ as a power series in x.

We have just seen that the spreads of call-of-mores are approximately proportional to the quantity *a*. It follows from this that the probability of profit from these transactions is independent of the time to maturity.

> *The probability of profit for the double call-of-more is* 0.394. *The transaction succeeds four times out of ten.*
> *The probability of profit for the triple call-of-more is* 0.33. *The transaction succeeds one time in three.*

The positive expectation of the call-of-more of order *n* is

$$n \int_r^\infty p(x - r)\, dx,$$

and as

$$\frac{r}{n-1} = \int_r^\infty p(x-r)\, dx,$$

the required expectation has value $(n/(n-1))r$, that is 1.36*a* for the double call-of-more and 1.64*a* for the triple call-of-more.

In selling a forward contract and simultaneously buying a double call-of-more, one obtains an option with premium $r = 0.68a$ and spread twice *r*. The probability of profit from the transaction is 0.30.

By analogy with options transactions, we call the transaction resulting from two call-of-mores of order *n* on upside and downside the *call-of-more straddle* of order *n*. The call-of-more straddle of second order is a very curious transaction: between the two prices $\pm r$ the loss is constant and equal to 2*r*. The loss diminishes progressively until the prices $\pm 3r$, where it vanishes. There is profit outside the interval $\pm 3r$. The probability of profit is 0.42.

FORWARD CONTRACTS

Now that we have completed the general study of probabilities we are going to apply it to the study of probabilities for the principal transactions of the exchange. We begin with the simplest, forward contracts and options, and we will finish with the study of combinations of these transactions.

The theory of speculation on commodities, much simpler than that on securities, has already been treated. We have, in fact, calculated the probability of profit from and the positive expectation of simple options, straddles and call-of-mores.

The theory of transactions of the exchange depends on two coefficients: *b* and *k*. Their value, at a given time, can easily be deduced from the spread of the forward contract from the spot price and from the spread of any option.

In the following study, we concentrate on the 3% bond, which is one of the securities on which options are regularly negotiated. We take for the values of b and k their average values over the last five years (1894 to 1898); that is to say,

$$b = 0.264, \qquad k = 5$$

(time is expressed in days and the unit of variation is the centime).

By *calculated* values, we mean those that are deduced from the formulae of the theory with the values above given to the constants b and k. The *observed* values are those that one deduces directly from the table of quoted prices during this same period from 1894 to 1898.[*]

In the chapters that follow we will constantly need to know the mean values of the quantity a at different times. The formula

$$a = 5\sqrt{t}$$

gives[50]

$$\text{for 45 days,} \quad a = 33.54,$$
$$\text{for 30 days,} \quad a = 27.38,$$
$$\text{for 20 days,} \quad a = 22.36,$$
$$\text{for 10 days,} \quad a = 16.13.$$

For one day, it seems that one must have $a = 5$, but in all the calculations of probabilities where one is concerned with averages one cannot put $t = 1$ for one day. In fact, there are 365 days in a year, but only 307 trading days.[51] The *average day* of the Exchange is therefore $t = \frac{365}{307}$; this gives

$$a = 5.45.$$

One can make the same remark for the coefficient b. In all the calculations relative to a day of the Exchange one must replace b by $b_1 = \frac{365}{307}b = 0.313$.

---

[*] *All the observations are taken from the 'Cote de la Bourse et de la Banque'.*

[50] The value for ten days should be 15.81. Many of the arithmetic values calculated by Bachelier are not terribly accurate.

[51] This is many more trading days than now. In 1900, the Exchange was open on Saturdays.

Probable Spread

Let us look for the interval of prices $(-\alpha, +\alpha)$ such that, at the end of the month, the bond will have the same chance of being inside the interval as of being outside. One must therefore have

$$\int_0^\alpha p \, dx = \tfrac{1}{4},$$

from which[52]

$$\alpha = \pm 46.$$

During the last sixty months, thirty-three times the variation has been contained within these limits and twenty-seven times they have been exceeded.

One can seek the same interval relative to one day. One then has

$$\alpha = \pm 9.$$

In 1452 observations, 815 times the variation has been less than 9 centimes.

In the preceding question, we have supposed that the quoted price coincides with the true price. Under these conditions, the probability of profit and the positive mathematical expectation of the buyer and the seller are the same. In reality, if n is the number of days from maturity, the quoted price is less than the true price by the quantity nb. The probable spread of 46 centimes on each side of the true price corresponds to the interval contained between 54 centimes above the quoted price and 38 centimes below this price.

Formula for the Probability in the General Case

To find the probability of an increase for a period of n days, we need to know the spread nb of the true price from the quoted price. The probability is then equal to

$$\int_{-nb}^\infty p \, dx.$$

The probability of a fall will be equal to one minus the probability of a rise.

---

[52]Recall that α is in units of centimes.

Probability of Profit from a Spot Purchase

We seek the probability of profit from a spot purchase that will be resold in thirty days. In the previous formula we must replace the quantity nb by 25. The probability is then equal to 0.64. The transaction has a two-in-three chance of success.

If one wants the probability for one year one must replace the quantity nb by 300. The formula $a = k\sqrt{t}$ gives

$$a = 95.5.$$

One finds that the probability is

$$0.89.$$

Nine times out of ten a spot purchase of government bonds produces a profit at the end of one year.

Probability of Profit from a Forward Purchase

We seek the probability of profit from a forward purchase made at the beginning of the month. One has

$$nb = 7.91, \qquad a = 27.38.$$

One deduces from this that

the probability of a rise is 0.55,

the probability of a fall is ...... 0.45.

The probability of profit from the purchase increases with time. For one year one has

$$n = 365, \qquad nb = 96.36, \qquad a = 95.5.$$

The probability then has value 0.65.

When one buys a forward contract to resell at the end of the year, one has a two-thirds chance of success. It is clear that if the monthly contango were 25 centimes, the probability of profit from the purchase would be 0.50.

Mathematical Advantage of Forward Transactions

As I have already said, it seems to me to be essential to study the mathematical advantage of a game as soon as it is not fair, and this is the case for forward contracts.

If we suppose that $b = 0$, the mathematical expectation of the forward purchase is $a - a = 0$. The advantage of the transaction is

$$\frac{a}{2a} = \frac{1}{2},$$

as in any fair game.

Let us seek the mathematical advantage of a forward purchase of n days, supposing that $b > 0$. During this period the buyer will have cashed the sum nb arising as the difference between the coupons and the contangoes and his expectation will be $a - a + nb$. His mathematical advantage will therefore be

$$\frac{a + nb}{2a + nb}.$$

The mathematical advantage of the seller will be

$$\frac{a}{2a + nb}.$$

Let us specialize to the case of the buyer. When $b > 0$ his mathematical advantage increases more and more with n. It is always greater than the probability of profit. For one month, the advantage of the buyer is 0.563 and his probability of profit is 0.55. For one year, his advantage is 0.667 and his probability of profit 0.65. One can therefore say:

The mathematical advantage of a forward transaction is almost equal to its probability of profit.

OPTIONS

Spread of Options

Knowing the value of a for a given time, one can easily calculate the true spread from the formula

$$m = \pi a \pm \sqrt{\pi^2 a^2 - 4\pi a(a - h)}.$$

Knowing the true spread, one obtains the quoted spread by adding the quantity nb to the true spread, where n is the number of days until maturity.

In the case of an option for next month, one adds the quantity $[25 + (n - 30)b]$.

Thus one arrives at the following results.

	Quoted Spread	
	Calculated	Observed
Options at 50 centimes		
At 45 days...	50.01	52.62
At 30 days...	20.69	21.22
At 20 days...	13.23	14.71
Options at 25 centimes		
At 45 days...	72.70	72.80
At 30 days...	37.78	37.84
At 20 days...	25.17	27.39
At 10 days...	12.24	17.40
Options at 10 centimes		
At 30 days...	66.19	60.93
At 20 days...	48.62	46.43
At 10 days...	26.91	32.89

In the case of an option at 5 centimes for the next day we have

$$h = 5, \qquad a = 5.45,$$

from which[53]

$$m = 0.81.$$

The true spread is therefore 5.81 and adding to it $b_1 = \frac{365}{307}b = 0.31$ one obtains the calculated spread 6.12. The average over the last five years gives 7.36.

The observed and calculated figures agree as a whole, but they show certain differences that must be explained.

Thus the observed spread of the option at 10 centimes at thirty days is too small. It is easy to understand the reason: in periods when there is a great deal of price movement, so that the option at 10 centimes would have a very large spread, this option is not quoted. The observed mean will therefore be diminished by this fact.

On the other hand, it cannot be denied that the market has had, for several years, a tendency to quote options with short maturities at too large spreads; it takes less account of the true proportion of smaller spreads with closer maturity. However, it must be added that it seems to have perceived its error because in 1898 it appears to have exaggerated in the opposite way.

---

[53] The displayed equation should read $m = 0.92$.

Probability of Exercise of Options

In order for an option to be exercised, the price at maturity must be bigger than the exercise price of the option. The probability of exercise is therefore given by the integral

$$\int_\epsilon^\infty p\,dx,$$

ϵ being the true exercise price of the option.

As we saw before, this integral is easy to calculate. It leads to the following results.

	Calculated	Observed
Probability of exercise of options at 50 centimes		
At 45 days...	0.63	0.59
At 30 days...	0.71	0.75
At 20 days...	0.77	0.76
Probability of exercise of options at 25 centimes		
At 45 days...	0.41	0.40
At 30 days...	0.47	0.46
At 20 days...	0.53	0.53
At 10 days...	0.65	0.65
Probability of exercise of options at 10 centimes		
At 30 days...	0.24	0.21
At 20 days...	0.28	0.26
At 10 days...	0.36	0.38

One can say that options at 50 centimes are exercised three times in four, the options at 25 centimes two times in four, and the option at 10 centimes one time in four.

The probability of exercise of the option at 5 centimes for the next day is calculated as 0.48. The result of 1456 observations gives 671 options certainly exercised and 76 for which the exercise is doubtful. Counting these last 76 options the probability would be 0.51, not counting them it would be 0.46, so on average 0.48, as predicted by the theory.

Probability of Profit from Options

In order for an option to give a profit to its buyer, the price at the call date must be greater than that of the option. The probability of profit is therefore expressible as the integral

$$\int_{\epsilon_1}^{\infty} p \, dx,$$

ϵ_1 being the price of the option.[54]
This integral leads to the following results.

	Calculated	Observed
Probability of profit from options at 50 centimes		
At 45 days...	0.40	0.39
At 30 days...	0.43	0.41
At 20 days...	0.44	0.40
Probability of profit from options at 25 centimes		
At 45 days...	0.30	0.27
At 30 days...	0.33	0.31
At 20 days...	0.36	0.30
At 10 days...	0.41	0.40
Probability of profit from options at 10 centimes		
At 30 days...	0.20	0.16
At 20 days...	0.22	0.18
At 10 days...	0.27	0.25

One sees that between ordinary practical limits, the probability of success of an option varies very little. The purchase at 50 centimes succeeds four times out of ten, the purchase at 25 centimes three times in ten and the purchase at 10 centimes two times in ten.

From the calculations, the buyer of an option at 5 centimes for the next day has a probability of success of 0.34. Observation of 1456 quotes shows that 410 options would certainly have given profits and 80 gave an uncertain result, the observed probability is therefore 0.31.

COMPLEX TRANSACTIONS

Classification of Complex Transactions

Since one negotiates forwards and often up to three options for the same settlement date, one can enter into triple and even quadruple transactions.

---

[54]Recall that in Bachelier's terminology the price is the exercise price plus the premium.

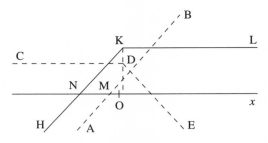

Figure 7.

The triple transactions already fall into a number of types that one can consider as classical; their study is very interesting but too long to be presented here. We therefore restrict ourselves to double transactions. One can divide these into two groups according to whether or not they contain a forward contract. The transactions that include a forward contract consist of the purchase of a forward and the sale of an option, or vice versa. The transactions of option against option consist of the sale of one big option followed by the purchase of a smaller one, or vice versa. The relative proportions of purchases and sales can moreover vary indefinitely. To simplify the question, we only study two very simple cases:

1. the second transaction is based on the same number of units as the first;
2. it is based on twice the number.

In order to fix ideas, we will suppose that transactions take place at the beginning of the month and we take for true spreads the average spread over the last five years: 12.78/50, 29.87/25 and 58.28/10. We also note that for transactions at one month the true price is higher than the quoted price by the quantity $7.91 = 30b$.

Forward Purchase against Sale of an Option

In reality one buys a forward contract at price $-30b = -7.91$ and sells an option at 25 centimes for 29.87. It is easy to represent the transaction geometrically (Figure 7). The forward purchase is represented by the straight line AMB. The length MO = $30b$. The sale of the option is represented by the broken line CDE. The resulting transaction will be represented by the broken line HNKL. The x-coordinate at the point N will be

$$-(25 + 30b).$$

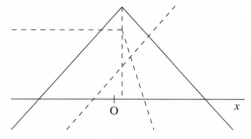

Figure 8.

One sees that the transaction gives a profit that is bounded above by the quoted spread of the option. The risk is unlimited below.

The probability of profit from the transaction is expressed by the integral

$$\int_{-25-30b}^{+\infty} p\,dx = 0.68.$$

If one had sold an option at 50 centimes the probability of success would have been 0.80.

It is interesting to know the probability in the case of a contango of 25 centimes ($b = 0$). This probability is 0.64 for selling an option at 25 centimes and 0.76 for selling an option at 50 centimes. If one sells an option against a spot purchase, the probability is 0.76 for selling at 25 and 0.86 for selling at 50.

Forward Sale against Purchase of an Option

This transaction is the inverse of the previous one. It gives a limited loss on the upside and an unlimited gain on the downside. It is, consequently, a downside option. For this option the spread is constant and the premium variable, the opposite of an upside option.

Forward Purchase against Sale of Two Options

One buys a forward at the true price $-30b$ and sells two options at 25 centimes at the price 29.87. Figure 8 represents the transaction geometrically. It shows that the risk is unlimited on the upside and on the downside.

One makes a profit between the prices $-(50 + 30b)$ and $59.74 + 30b$. The probability of profit is

$$\int p\,dx = 0.64.$$

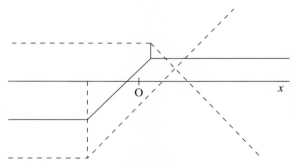

Figure 9.

On selling at 50 centimes the probability would be 0.62 and on selling at 10 centimes one would have probability 0.62 of profit. If one had bought two forwards and sold three options at 50, the probability would have been 0.66.

Forward Sale against Purchase of Two Options

This is the inverse transaction of the previous one. It gives a profit in the case of a big rise and in the case of a big fall. Its probability of profit is 0.27.

Purchase of a Very Large Option against Sale of a Small One

I suppose that one has simultaneously agreed the following two transactions:

$$\text{purchase at} \ldots \ldots 12.78/50,$$
$$\text{sale at} \ldots \ldots 29.87/25.$$

Below the exercise price of the big option (-37.22), the two options are abandoned and one loses 25 centimes. Starting from price -37.22 one is a buyer, and at price -12.22 the transaction has zero value. One gains from then until the exercise price of the option at 25, that is to say when the price $+4.87$ is reached. Then the transactions are settled and one gains the spread. On the downside one loses 25 centimes, this is the maximum risk. On the upside one gains the spread. The risk is limited, as is the profit. Figure 9 is a geometric representation of the transaction.

The probability of profit is given by the integral

$$\int_{-12.22}^{\infty} p \, dx = 0.59.$$

Buying an option at 25 and selling one at 10, the probability of profit would be 0.38.

Sale of a Big Option against the Purchase of a Small One

This transaction, which is the counterpart of the previous one, is discussed without difficulty. On the downside one gains the difference of the amounts of the options, on the upside one loses their spread.

Purchase of a Large Option against Sale of Two Small Options

I suppose that one has made the following transactions:

buy an option at 12.78/50,

sell two options at . . . . . . 29.87/25.

If there is a big rise in prices the options are abandoned and the premiums cancel; it is a break-even transaction. At the exercise price of the big option, that is to say at price -37.22, one becomes a buyer and gains progressively up until the exercise price of the small one ($+4.87$). At this point, the profit is maximized (42.09 centimes) and one becomes a seller. The profit decreases progressively and at the price 45.96 it vanishes. After this the loss is proportional to the increase in price. In summary, the transaction gives a limited gain, a zero risk on the downside and an unlimited loss on the upside. Figure 10 represents this transaction geometrically.

Sale of a Large Option against the Purchase of Two Small Options

The discussion and geometric representation of this transaction, the inverse of the preceding one, presents no difficulty. It is unnecessary for us to do it here.

Practical Classification of the Transactions of the Exchange

From the practical point of view, one can divide the transactions of the exchange into four classes:
- upside transactions;
- downside transactions;
- transactions in anticipation of a big price movement in some direction;
- transactions in anticipation of small price movements.

61

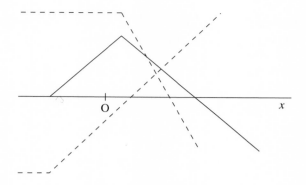

Probability of breaking even......0.30
Probability of making a profit......0.45
Probability of making a loss......0.25

Figure 10.

The following table summarizes the principal upside transactions.

	Average probability		
	$b = \frac{25}{30}$ (zero contango)	$b = 0.26$ (average contango)	$b = 0$ (contango = coupon)
Purchase/10	0.20	0.20	0.20
Purchase/25	0.33	0.33	0.33
Purchase/25 against sale/10	0.38	0.38	0.38
Purchase/50	0.43	0.43	0.43
Forward purchase	0.64	0.55	0.50
Purchase/50 against sale/25	0.59	0.59	0.59
Forward purchase against sale/25	0.76	0.68	0.64
Forward purchase against sale/50	0.86	0.80	0.76

To obtain the scale of downside transactions it suffices to invert this table.

PROBABILITY THAT A PRICE WILL BE ATTAINED IN A GIVEN INTERVAL OF TIME

Let us seek the probability P that a given price c will be attained or exceeded in an interval of time t.

Suppose, first of all, for simplicity, that the time is decomposed in two units: that t is two days, for example. Let x be the price quoted on the first day and y the price on the second day relative to that of the first. In order for the price c to be attained or exceeded, it is necessary that on the first day the price is contained between c and ∞ or that on the second day it is contained between $c - x$ and ∞. We must distinguish four cases:

First day	Second day
x between	y between
$-\infty$ and c	$-\infty$ and $c - x$
$-\infty$ and c	$c - x$ and ∞
c and ∞	$-\infty$ and $c - x$
c and ∞	$c - x$ and ∞

Of these four cases, the last three are favourable.

The probability that the price is in the interval dx the first day and dy the second day will be $p_x p_y \, dx \, dy$. The probability P, being by definition the ratio of the number of favourable cases to all possible cases, will be

$$P = \frac{\int_{-\infty}^{c} \int_{c-x}^{\infty} + \int_{c}^{\infty} \int_{-\infty}^{c-x} + \int_{c}^{\infty} \int_{c-x}^{\infty}}{\int_{-\infty}^{c} \int_{-\infty}^{c-x} + \int_{-\infty}^{c} \int_{c-x}^{\infty} + \int_{c}^{\infty} \int_{-\infty}^{c-x} + \int_{c}^{\infty} \int_{c-x}^{\infty}}$$

(the integrand is $p_x p_y \, dx \, dy$).

The four integrals in the denominator represent the four possible cases, the three integrals in the numerator represent the favourable cases. Since the denominator is equal to one, one can simplify and write

$$P = \int_{-\infty}^{c} \int_{c-x}^{\infty} p_x p_y \, dx \, dy + \int_{c}^{\infty} \int_{-\infty}^{\infty} p_x p_y \, dx \, dy.$$

One can apply the same reasoning to the case when there are three consecutive days to consider, then four, etc. This method leads to more and more complicated expressions because the number of favourable cases will increase without bound. It is much simpler to study the probability $1 - P$ that the price c will never be attained. There is then no more than one favourable case for any number of days, namely that where the price is not attained on any of the days under consideration.

The probability $1 - P$ is expressed as

$$1 - P = \int_{-\infty}^{c} \int_{-\infty}^{c-x_1} \int_{-\infty}^{c-x_1-x_2} \cdots \int_{-\infty}^{c-x_1-x_2-\cdots-x_{n-1}} p_{x_1} \cdots p_{x_n} \, dx_1 \cdots dx_n,$$

where x_1 is the price on the first day, x_2 is the price on the second day relative to that on the first, x_3 is the relative price on the third day, etc.

63

The evaluation of this integral appears difficult; we resolve the question using an approximation method.

One can consider the time t as divided into small intervals Δt in such a way that $t = m\Delta t$. During the unit of time Δt, the price only varies by the quantity $\pm \Delta x$, the mean spread relative to this unit of time. Each of the spreads $\pm \Delta x$ has probability $\frac{1}{2}$.

Let us suppose that $c = n\Delta x$ and let us seek the probability that the price c is attained precisely at time t, that is to say that the price c is attained at this time t without ever having been attained before. If, during the m time units, the price has changed by the quantity $n\Delta x$, that is because there have been $\frac{1}{2}(m + n)$ upward movements and $\frac{1}{2}(m - n)$ downward movements. The probability that in m variations there have been $\frac{1}{2}(m + n)$ favourable ones is

$$\frac{m!}{\frac{1}{2}(m-n)!\frac{1}{2}(m+n)!}(\tfrac{1}{2})^m.$$

It is not this probability that we are looking for, but the product of this probability with the ratio of the number of ways in which the price $n\Delta x$ is attained at the time $m\Delta t$, without having been attained before, to the total number of ways in which it is attained at time $m\Delta t$. We will calculate this ratio.

During the m units of time that we are considering, there have been $\frac{1}{2}(m + n)$ upward movements and $\frac{1}{2}(m - n)$ downward movements. We can represent one of the combinations giving a rise of $n\Delta x$ in m time units by the expression

$$B_1 H_1 H_2 \cdots B_{(m-n)/2} \cdots H_{(m+n)/2}.$$

B_1 indicates that, during the first time unit, there was a fall. H_1, which comes next, indicates that there was a rise during the second unit of time, etc. In order for a combination to be favourable, it must be the case that, reading from right to left, the number of H's is always greater than the number of B's. We have arrived at the following problem.

> In n letters there are $\frac{1}{2}(m + n)$ H's and $\frac{1}{2}(m - n)$ B's; what is the probability that, writing the letters randomly and reading in an agreed direction, the number of H's will, throughout the reading, always be greater than the number of B's?

The solution to this problem, presented in a slightly different form, has been given by M. André. The required probability is equal to n/m.

The probability that the price $n\Delta x$ will be attained precisely at the end of m time units is therefore

$$\frac{n}{m}\frac{m!}{\frac{1}{2}(m-n)!\frac{1}{2}(m+n)!}(\tfrac{1}{2})^m.$$

This formula is approximate. We would obtain a more exact expression by replacing the quantity that multiplies n/m by the exact value of the probability at time t, that is to say by[55]

$$\frac{\sqrt{2}}{\sqrt{m}\sqrt{\pi}}\exp\left[-\frac{n^2}{\pi m}\right].$$

The probability that we are looking for is therefore

$$\frac{n\sqrt{2}}{m\sqrt{m}\sqrt{\pi}}\exp\left[-\frac{n^2}{\pi m}\right],$$

or, replacing n by $2c\sqrt{\pi}/\sqrt{2}$ and m by $8\pi k^2 t$,

$$\frac{\mathrm{d}t c\sqrt{2}}{2\sqrt{\pi}kt\sqrt{t}}\exp\left[-\frac{c^2}{4\pi k^2 t}\right].$$

This is the expression for the probability that the price c is attained at time $\mathrm{d}t$, never having been attained before.

The probability that the price c will not be attained before time t has value

$$1-P = A\int_t^\infty \frac{c\sqrt{2}}{2\sqrt{\pi}kt\sqrt{t}}\exp\left[-\frac{c^2}{4\pi k^2 t}\right]\mathrm{d}t.$$

I have multiplied the integral by a constant A to be determined, because the price cannot be attained unless the quantity denoted by m is even.

---

[55]Bachelier perpetuates his previous errors here. In fact the probability that we are looking for is

$$\frac{n\sqrt{2}}{m\sqrt{m}\sqrt{\pi}}\exp\left[-\frac{2n^2}{m}\right],$$

and using $(\Delta x)^2/\Delta t = 2\pi k^2$, $2n\Delta x = c$ and $m\Delta t = t$ we obtain

$$\frac{c\Delta t}{2\pi t\sqrt{t}k}\exp\left[-\frac{c^2}{4\pi k^2 t}\right].$$

Notice that in the notation that Bachelier now introduces this is

$$\frac{\Delta t}{\mathrm{d}t}\frac{Ac\sqrt{2}}{2\sqrt{\pi}kt\sqrt{t}}\exp\left[-\frac{c^2}{4\pi k^2 t}\right]\mathrm{d}t.$$

His reasoning for introducing the constant A is fallacious, but this trick does result in him recovering from his error and eventually obtaining the correct formula for the probability P.

Putting

$$\lambda^2 = \frac{c^2}{4\pi k^2 t},$$

one has

$$1 - P = 2\sqrt{2}A \int_0^{c/(2\sqrt{\pi}k\sqrt{t})} e^{-\lambda^2}\, d\lambda.$$

To determine A, we put $c = \infty$, then $P = 0$ and

$$1 = 2\sqrt{2}A \int_0^{\infty} e^{-\lambda^2}\, d\lambda = \sqrt{2}\sqrt{\pi}A.$$

Therefore,

$$A = \frac{1}{\sqrt{2}\sqrt{\pi}}$$

and thus

$$1 - P = \frac{2}{\sqrt{\pi}} \int_0^{c/(2\sqrt{\pi}k\sqrt{t})} e^{-\lambda^2}\, d\lambda.$$

The probability that the price x will be attained or exceeded during the interval of time up to t is therefore

$$P = 1 - \frac{2}{\sqrt{\pi}} \int_0^{x/(2\sqrt{\pi}k\sqrt{t})} e^{-\lambda^2}\, d\lambda.$$

The probability that the price x will be attained or exceeded *at time t* is, as we have seen,

$$\mathcal{P} = \frac{1}{2} - \frac{1}{2}\frac{2}{\sqrt{\pi}} \int_0^{x/(2\sqrt{\pi}k\sqrt{t})} e^{-\lambda^2}\, d\lambda.$$

One sees that $\mathcal{P}$ is half of P.

The probability that a price will be attained or exceeded at time t is half the probability that the price will be attained or exceeded during the interval of time up to t.

The direct proof of this result is very simple: the price cannot be exceeded at time t without having been attained previously. The probability $\mathcal{P}$ is therefore equal to the probability P multiplied by the probability that the price is exceeded at time t given that it was quoted at a time prior to t, that is to say multiplied by $\frac{1}{2}$. One therefore has[56]

$$\mathcal{P} = \tfrac{1}{2}P.$$

---

[56]This result is now known as the *reflection principle*. It is usually attributed to Désiré André, who (as credited by Bachelier) proved an analogous result (in purely combinatorial form) in a different context. Bachelier's first argument uses the construction of Brownian motion as a limit of random walks. To be made rigorous, the second, which receives strong praise from Poincaré in his report, requires the *strong Markov property*

One can see that the multiple integral that expresses the probability $1 - P$ and appears to resist ordinary approaches of calculus is determined by a very elementary argument as a result of the calculus of probabilities.

Applications

The tables of the function Θ permit very simple calculation of the probability

$$P = 1 - \Theta\left(\frac{x}{2\sqrt{\pi}k\sqrt{t}}\right).$$

The equation

$$P = 1 - \frac{2}{\sqrt{\pi}} \int_0^{x/(2\sqrt{\pi}k\sqrt{t})} e^{-\lambda^2}\, d\lambda$$

shows that the probability is constant when the spread x is proportional to the square root of time; that is to say, when there is an expression of the form $x = ma$. We are going to study the probabilities corresponding to certain interesting spreads.

Let us suppose first of all that $x = a = k\sqrt{t}$; the probability P is then equal to 0.69. When the spread a is attained, one can, without loss, resell a forward contract against the simple option a.[57] Therefore,

> *There are two chances in three that one can, without loss, resell a forward contract against a simple option.*

Let us specialize the question to the 3% bond. In thirty-eight out of sixty months observed it has been possible to resell a forward contract at spread a, which corresponds to a probability of 0.63.

Let us now study the case when $x = 2a$. The preceding formula gives probability 0.43. When the spread $2a$ is attained, one can resell without

---

for Brownian motion—an extension of the lack-of-memory property of the process to certain random times. This property was properly formulated by Doob, who proved it for processes with a countable state space in 1945. It was not until 1956 that Ray showed that the Markov property does not imply the strong Markov property and that Brownian motion does indeed satisfy the strong Markov property. Symmetry arguments of this type were employed to great effect by Lévy in his studies of Brownian motion (see, for example, Lévy 1948).

[57] That is, the holder of the simple option can sell a forward contract to lock in a profit. If the price continues to rise, then the loss on the sale offsets any additional profit on the option, so that the profit is never more than a. If, on the other hand, the price falls, the profit on the sale offsets the loss on the option. The loss on the option is limited to the premium, whereas the profit on the sale of the forward contract is unlimited.

loss a forward contract against an option with premium $2a$. Thus,

> There are four chances in ten that one can, without loss, resell a forward contract against an option with premium $2a$.

In a period of sixty liquidation dates, the 3% bond attained the spread $2a$ twenty-three times, which gives probability 0.38.

The spread $0.7a$ is that of the call-of-more of order two. The corresponding probability is 0.78.

> One has three chances in four of being able to, without loss, resell a forward against a call-of-more of order two.

The call-of-more of order three must be traded at a spread of $1.1a$, which corresponds to the probability 0.66.

> One has two chances in three of being able to, without loss, resell a forward against a call-of-more of order three.

Among notable spreads, we finally mention $1.7a$, which corresponds to a probability of $\frac{1}{2}$, and the spread of $2.9a$, which corresponds to a probability of $\frac{1}{4}$.

Apparent Mathematical Expectation

The mathematical expectation

$$\mathcal{E}_1 = Px = x - \frac{2x}{\sqrt{\pi}} \int_0^{x/(2\sqrt{\pi}k\sqrt{t})} e^{-\lambda^2} \, d\lambda$$

is a function of x and t. Differentiating with respect to x we obtain

$$\frac{\partial \mathcal{E}_1}{\partial x} = 1 - \frac{2}{\sqrt{\pi}} \int_0^{x/(2\sqrt{\pi}k\sqrt{t})} e^{-\lambda^2} \, d\lambda - \frac{xe^{-x^2/(4\pi k^2 t)}}{\pi k \sqrt{t}}.$$

If one considers a fixed time t, this expectation will be maximized when

$$\frac{\partial \mathcal{E}_1}{\partial x} = 0,$$

that is to say approximately when $x = 2a$.

Apparent Total Expectation

The total expectation corresponding to time t will be the integral

$$\int_0^\infty Px \, dx.$$

Let us put

$$f(a) = \int_0^\infty \left(x - \frac{2x}{\sqrt{\pi}} \int_0^{x/(2\sqrt{\pi}k\sqrt{t})} e^{-\lambda^2} \, d\lambda \right) dx.$$

Differentiating with respect to a, we obtain[58]

$$f'(a) = \frac{1}{\pi a^2} \int_0^\infty x^2 \exp\left[-\frac{x^2}{4\pi a^2} \right] dx,$$

or $f'(a) = 2\pi a$. Thus

$$f(a) = \pi a^2 = \pi k^2 t.$$

The total expectation is proportional to time.[59]

Time of the Greatest Probability

The probability

$$P = 1 - \frac{2}{\sqrt{\pi}} \int_0^{x/(2\sqrt{\pi}k\sqrt{t})} e^{-\lambda^2} \, d\lambda$$

is a function of x and t. The study of how it changes as x varies presents nothing of particular note: the function decreases constantly as x increases.

Let us suppose now that x is constant and study the variation of P as a function of t. Differentiating, we obtain[60]

$$\frac{\partial P}{\partial t} = \frac{x e^{-x^2/(4\pi k^2 t)}}{2\pi t \sqrt{t}}.$$

We determine the time when the probability of hitting level x for the first time is maximized by setting the derivative equal to zero:

$$\frac{\partial^2 P}{\partial t^2} = \frac{x e^{-x^2/(4\pi k^2 t)}}{2\pi k t \sqrt{t}} \left(\frac{x^2}{4\pi k^2 t} - \frac{3}{2} \right),$$

which gives

$$t = \frac{x^2}{6\pi k^2}.$$

Suppose, for example, that $x = k\sqrt{t_1}$, we will then have

$$t = \frac{t_1}{6\pi}.$$

---

[58]The dependence of the integral defining $f(a)$ on a is through the coefficient of instability. Recall that $a = k\sqrt{t}$.

[59]Bachelier does not interpret this for us. It tells us about the second moment of the maximum of the price up to time t.

[60]There is a k missing from the denominator of the fraction on the right-hand side.

The most probable time at which one can without loss resell the forward contract against a simple option is at one-eighteenth of the duration of the maturity.

If we now suppose that $x = 2k\sqrt{t}$, we obtain $t = 2t_1/3\pi$.

The most probable time at which one can without loss resell the forward contract against an option with premium $2a$ is at one-fifth of the duration of the maturity.

The probability P corresponding to the time $t = x^2/6\pi k^2$ has value $1 - \Theta(\frac{1}{2}\sqrt{6}) = 0.08$.

Average Time

When an event can happen at different times, the sum of the products of the possible durations and their respective probabilities is called the average arrival time of the event. The average duration is equal to the sum of the expectations of the possible durations.[61] The average time at which the price x will be exceeded is therefore expressed by the integral

$$\int_0^\infty t \frac{dP}{dt}\, dt = \int_0^\infty \frac{x}{2\pi k\sqrt{t}} \exp\left[- \frac{x^2}{4\pi k^2 t} \right] dt.$$

Putting $x^2/4\pi k^2 t = y^2$ this becomes

$$\frac{x^2}{2\pi\sqrt{\pi}k^2} \int_0^\infty \frac{e^{-y^2}}{y^2}\, dy.$$

This integral is infinite. The average time is therefore infinite.

Absolute Probable Instant[62]

This will be the time at which one has $P = \frac{1}{2}$ or

$$\Theta\left(\frac{x}{2\sqrt{\pi}k\sqrt{t}} \right) = \frac{1}{2};$$

from which one deduces

$$t = \frac{x^2}{2.89k^2}.$$

---

[61] Recall that for Bachelier the expectation of a possible duration is its value multiplied by the probability of that duration, whereas in modern probability theory we use the term expectation to mean the sum of these quantities over all possible durations. Bachelier's 'average duration' corresponds to the modern notion of 'expected duration'.

[62] Whereas the time of the greatest probability found the most likely time for a given level to be attained by the price for the first time, the absolute probable instant is the time at which the probability that the level has been exceeded first exceeds $\frac{1}{2}$.

The absolute probable instant varies, in the same way as the most probable instant, in proportion to the square of the quantity x, and it is about six times bigger than the most probable instant.

Relative Probable Instant

It is interesting to know not only the probability that a price x will be quoted in an interval of time t, but also the probable time T at which this price will be attained. This time is evidently different from that with which we have just been concerned. The interval of time T will be such that there are equal chances that the price will be quoted before the time T as that it will be quoted afterwards, that is to say in the interval T, t.[63]
T will be given by the formula

$$\int_0^T \frac{\partial P}{\partial t}\,dt = \frac{1}{2}\int_0^t \frac{\partial P}{\partial t}\,dt$$

or

$$1 - 2\Theta\left(\frac{x}{2\sqrt{\pi}k\sqrt{T}}\right) = -\Theta\left(\frac{x}{2\sqrt{\pi}k\sqrt{t}}\right).$$

As an application of this, suppose that $x = k\sqrt{t}$; the formula gives $T = 0.18t$. Therefore,

> One has as much chance of being able to resell without loss a forward contract against a simple option during the first fifth of the duration of the contract as during the other four-fifths.

To treat a particular example, suppose that we are concerned with bonds and that $t = 30$ days, then T will equal 5 days. There is therefore as much chance, we learn from the formula, that one can resell the bond with spread a (28 centimes on average) during the first five days as the chance that one can resell during the twenty-five days that follow. During the sixty liquidations that we observed, the spread was attained thirty-eight times: eighteen times during the first four days, twice during the fifth and eighteen times after the fifth day. The observation is therefore in agreement with the theory.

Suppose now that $x = 2k\sqrt{t}$, we find $T = 0.42t$. As the quantity $2k\sqrt{t}$ is the spread of the option with premium $2a$, one can therefore say,

---

[63] In other words, we are looking at a *conditional* probability here. Given that the price has been attained by time t, if T is the relative probable instant, then the probability that the price was first quoted in $[0, T]$ is the same as the probability that it was first quoted in $[T, t]$.

There is as much chance that one can without loss resell the forward against an option with premium 2a during the first four-tenths of the duration of the contract as during the other six-tenths.

Let us turn once again to government bonds. Our previous observations show us that in twenty-three cases out of sixty liquidations the spread 2a (56 centimes on average) was attained. Of the twenty-three cases, the spread was attained before the fourteenth of the month eleven times and after this time twelve times.

The probable time will be $0.11t$ for the call-of-more of order two and $0.21t$ for the call-of-more of order three.

Finally, the probable instant will be half of the total time if x is equal to $2.5k\sqrt{t}$.

Distribution of the Probability

Up until now we have solved two problems:

- the probability of a given price being attained at time t;
- the probability that a price will be attained in a time interval of length t.

We will now resolve this last problem completely. It is not sufficient to know the probability that a price will be attained before a time t, it is also necessary to know the law of probability at time t in the case when the price has not been attained. I suppose, for example, that we buy a government bond to resell with a profit c. If it has not been possible to effect the resale by time t, what will the probability law of our transaction be at this time?

If the price c has not been attained, this is because the maximum upward variation of the price has been less than c, while the downward variation could have been indefinite. There is therefore an evident asymmetry in the curve of probabilities at time t. We seek the form of this curve.

Let ABCEG be the curve of probabilities at time t, supposing that the transaction will persist until this time (Figure 11).

The probability that, at time t, the price c will be exceeded is represented by the area DCEG, which, evidently, will no longer form part of the curve of probabilities in the case when resale is possible.

We can even assert *a priori* that the area under the curve of probabilities must again, in this case, be diminished by a quantity equal to DCEG, since the probability P is twice the probability represented by DCEG.

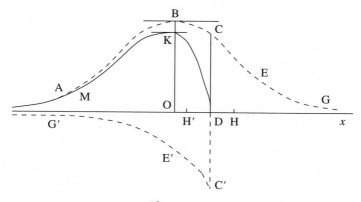

Figure 11.

If the price c is attained at time t_1, the price H will have, at this instant, the same probability as the symmetric price H'. The probability of resale at price c includes not only the probability of H but also an equal probability of H', and so for the probability at time t we must subtract from the y-coordinates of the curve ABC those of the curve $G'E'C'$, symmetric to GEC. The curve of probabilities that we seek will therefore be the curve DKM.

This curve has equation[64]

$$p = \frac{1}{2\pi k\sqrt{t}}\left[\exp\left[-\frac{x^2}{4\pi k^2 t}\right] - \exp\left[-\frac{(2c-x)^2}{4\pi k^2 t}\right]\right].$$

Curve of the Maximum Probability

To obtain the price for which the probability is greatest, in the case when the price c has not been attained, it is enough to set $dp/dx = 0$; in this way one obtains

$$\frac{x}{2c-x} + \exp\left[-\frac{c(c-x)}{\pi k^2 t}\right] = 0.$$

If one supposes that $c = a = k\sqrt{t}$, one obtains

$$x_m = -1.5a;$$

---

[64]Effectively Bachelier is using Lord Kelvin's method of images for calculating the fundamental solution to the heat equation with an absorbing boundary condition. Of course he has not obtained a probability distribution. For that he needs to divide this expression by the probability that the price c has indeed not been attained by time t.

if one supposes that $c = 2a$, one obtains[65]

$$x_m = -0.4a.$$

Finally, one would obtain

$$x_m = -c$$

if c were equal to $1.33a$.

Probable Price

We seek the expression for the probability that the price is in the interval $0, u$. This will be

$$\frac{1}{2\pi k\sqrt{t}} \int_0^u \exp\left[-\frac{x^2}{4\pi k^2 t}\right] dx - \frac{1}{2\pi k\sqrt{t}} \int_0^u \exp\left[-\frac{(2c-x)^2}{4\pi k^2 t}\right] dx.$$

The first term has value

$$\tfrac{1}{2}\Theta\left(\frac{u}{2\sqrt{\pi}k\sqrt{t}}\right).$$

For the second, putting

$$2\sqrt{\pi}k\sqrt{t}\lambda = 2c - x,$$

this term will become

$$-\frac{1}{2}\frac{2}{\sqrt{\pi}} \int_0^{2c/(2\sqrt{\pi}k\sqrt{t})} e^{-\lambda^2} \, d\lambda + \frac{1}{2}\frac{2}{\sqrt{\pi}} \int_0^{(2c-u)/(2\sqrt{\pi}k\sqrt{t})} e^{-\lambda^2} \, d\lambda.$$

The expression for the probability is therefore

$$\tfrac{1}{2}\Theta\left(\frac{u}{2\sqrt{\pi}k\sqrt{t}}\right) - \tfrac{1}{2}\Theta\left(\frac{2c}{2\sqrt{\pi}k\sqrt{t}}\right) + \tfrac{1}{2}\Theta\left(\frac{2c-u}{2\sqrt{\pi}k\sqrt{t}}\right).$$

It is interesting to study the case when $u = c$ to find out the probability of profit from purchase of a forward contract when the resale price has not been attained.

Under the hypothesis $u = c$, the formula above becomes

$$\Theta\left(\frac{c}{2\sqrt{\pi}k\sqrt{t}}\right) - \tfrac{1}{2}\Theta\left(\frac{2c}{2\sqrt{\pi}k\sqrt{t}}\right).$$

Supposing that $c = a$, the probability is then 0.03. If the spread a has never been attained in the interval up to t, there are no more than three chances in one hundred that, at time t, the price is between 0 and a.[66]

---

[65] This should be $-0.8a$.

[66] This, and the corresponding claim for the spread of $2a$, is misleading. Bachelier is not actually calculating the conditional probability. What he has actually shown is that there is no more than a 3% chance that the spread a will not be attained by time t *and* at time t the price is between 0 and a.

One can buy a simple option with the preconceived idea of resale of the forward contract against this option when its spread is attained. The probability of resale is, as we have seen, 0.69. The probability that the resale has not taken place and that there is profit is 0.03 and the probability of loss is 0.28.

Let us suppose that $c = 2a$, the probability is then 0.13. If the spread $2a$ has never been attained in the interval t, there are thirteen chances in one hundred that, at time t, the price will be between 0 and $2a$.

The probable curve is that for which the ordinate divides the area under the curve of probabilities into two equal parts. It is impossible to express its value in finite terms.

Real Expectation

The mathematical expectation $k\sqrt{t} = a$ expresses the expectation of a transaction that must last until time t. If one intends to realize the transaction if a certain spread is attained before time t, the mathematical expectation has a completely different value, varying evidently between zero and $k\sqrt{t}$ as the chosen spread varies between zero and infinity.

For example, let c be the price at which a purchase will be realized. To obtain the real positive expectation, one must add to the expectation of resale cP the positive expectation corresponding to the case when the resale has not happened, that is to say the quantity

$$\int_0^c \frac{x}{2\pi k\sqrt{t}}\left[\exp\left[-\frac{x^2}{4\pi k^2 t} \right] - \exp\left[-\frac{(2c-x)^2}{4\pi k^2 t} \right] \right] dx.$$

If one performs the integration of the first term and adds the whole integral to the expectation of resale,

$$cP = c - c\frac{2}{\sqrt{\pi}} \int_0^{c/(2\sqrt{\pi}k\sqrt{t})} e^{-\lambda^2}\, d\lambda,$$

one obtains an expression for the real expectation:

$$\mathcal{E} = c + k\sqrt{t}\left(1 - \exp\left[-\frac{c^2}{\pi k^2 t} \right]\right) - c\frac{2}{\sqrt{\pi}}\int_0^{c/(\sqrt{\pi}k\sqrt{t})} e^{-\lambda^2}\, d\lambda$$

or

$$\mathcal{E} = c + k\sqrt{t}\left(1 - \exp\left[-\frac{c^2}{\pi k^2 t} \right]\right) - c\Theta\left(\frac{c}{\sqrt{\pi}k\sqrt{t}} \right).$$

75

If we suppose that $c = \infty$, we find again that $\mathcal{E} = k\sqrt{t}$. One can easily develop $\mathcal{E}$ in a series, but the preceding formula is more useful. It is calculated with tables of logarithms and of the function Θ.

For $c = a$ one obtains

$$\mathcal{E} = 0.71a.$$

Likewise, for $c = 2a$ one has

$$\mathcal{E} = 0.95a,$$

the expectations of resale being, for these same spreads, $0.69a$ and $0.86a$.

The *average spread* on the downside, when the price c is not attained, has value

$$\frac{\int_{-\infty}^{0} px \, dx}{\int_{-\infty}^{0} p \, dx} = \frac{\mathcal{E}}{1 - P - P_1},$$

where P_1 denotes the quantity $\int_0^c p \, dx$. The mean spread therefore has value $2.54a$ when $c = a$ and $2.16a$ when $c = 2a$.

If one supposes that $c = \infty$, one sees that the average spread is equal to $2a$, a result obtained before.

Let us take again, as an example, the general problem relating to the spread a. I buy a forward contract with the intention of reselling with the spread $a = k\sqrt{t}$. If at time t the sale has not been effected, I sell at whatever the price. What are the principal results that the calculus of probabilities furnishes for this transaction?

- The real positive expectation of the transaction is $0.71a$.
- The probability of resale is 0.69.
- The most probable time of resale is $\frac{1}{18}t$.
- The probable time[67] of resale is $\frac{1}{5}t$.
- If the resale has not taken place, the probability of success is 0.03, the probability of loss is 0.28, the positive expectation is $0.02a$, the negative expectation is $0.71a$, the average loss is $2.54a$.
- The total probability of profit is 0.72.

I do not believe that it is necessary to present other examples; one sees that the current theory resolves by the calculus of probabilities most of the problems to which the study of speculation leads.

A last remark will perhaps not be unprofitable. If in consideration of several questions treated in this study I have compared the results of

---

[67]Bachelier previously called this the 'relative probable instant'.

observation to those of the theory, this was not to verify the formulae established by mathematical methods, but only to show that the market, unwittingly, obeys a law that rules it: the law of probability.

Seen and approved:

Paris, 6 January 1900.
Dean of the faculty of sciences,
G. DARBOUX

Seen and passed for publication:

Paris, 6 January 1900.
The vice-rector of the Académie de Paris,
GRÉARD

REPORT ON BACHELIER'S THESIS (29 MARCH 1900)

Appell, Poincaré, J Boussinesq

The subject chosen by M. Bachelier is rather far away from those usually treated by our candidates; his thesis is entitled 'Theory of speculation' and its subject is the application of the calculus of probabilities to transactions of the exchange. At first one might have feared that the author allowed himself to be deluded about the range of the calculus of probabilities, as has often happened. Happily this is not the case; in his introduction and later in the paragraph entitled the probability in the transactions of the exchange he does his utmost to fix the limits within which one can have legitimate recourse to this type of calculus; he does not therefore exaggerate the range of his results and I do not believe that he is deceived by his formulae.

What then can one legitimately assert in this subject? It is clear first of all that the relative prices of different types of transactions must obey certain laws; thus one could imagine combinations of prices on which one could bet with certainty. The author cites some examples. It is clear that such combinations are never produced, or that if they are produced they will not persist. The buyer believes in a probable rise, without which he would not buy, but if he buys, there is someone who sells to him and this seller believes a fall to be probable. From this it follows that the market taken as a whole considers the mathematical expectation of all transactions and all combinations of transactions to be null.

What are the mathematical consequences of such a principle? If one assumes that the deviations are not too great, one can suppose that the probability of a given deviation from the quoted price does not depend on the absolute value of this price. Under these conditions the principle of the mathematical expectation is sufficient to recover the law of the probabilities; one falls again on the celebrated law of errors of Gauss.

As this law has been the subject of numerous proofs that in the main are simple paralogisms, it is advisable to be circumspect and to examine this proof closely; or at least it is necessary to set out in a precise manner the hypotheses that one is making. Here the hypothesis that one has to make is, as I have just said, that the probability of a given deviation from the current price is independent of the absolute value of the price. The hypothesis can be accepted provided that the deviations are not too big. The author sets this out clearly without perhaps stressing it as much as would be advisable. It is enough however that he has written it explicitly for his arguments to be correct.

The manner in which M. Bachelier extracts Gauss's law is very original and all the more interesting as his reasoning could extend with some changes to the theory of errors itself. He develops it in a chapter whose title seems a little strange, because he calls it 'Radiation of Probability'. It is in effect a comparison with the analytic theory of heat propagation to which the author has had recourse. A little reflection shows that the analogy is real and the comparison legitimate. The reasoning of Fourier is applicable almost without change to this problem, so different from that for which it was created.

One can regret that M. Bachelier has not developed further this part of his thesis. He could have entered into the details of Fourier's analysis. He has said enough about it however to justify the Gauss law and to catch a glimpse of the case where it ceases to be legitimate.

The Gauss law having been established, one can deduce quite easily from it certain consequences that can be verified experimentally. One such example is the relation between the value of an option and the spread with the forward. One must not expect too exact a verification. The principle of mathematical expectation imposes itself in the sense that, if it were violated, there would always be people who had an interest in playing in such a way as to reestablish it and they would end up seeing it. But they will not see it unless the gap is considerable. The verification cannot then be more than coarse. The author of the thesis gives statistics where the verification is made in a very satisfying manner.

M. Bachelier next examines a problem which at first appears to require the use of very complicated calculus. What is the probability that a price will be attained before some date? In writing the equation for the problem one is led to a multiple integral in which one sees as many superposed integral signs as there are days before the fixed date. This equation seems at first to be inaccessible. The author resolves it with a short, simple and elegant argument; he moreover notes the analogy with the known reasoning of M André for the ballot problem. But this analogy is not strict enough to diminish in any way the originality of this ingenious trick. For other analogous problems the author applies it with equal success.

In summary, we are of the opinion that there is reason to authorize M. Bachelier to print his thesis and to submit it.

From Bachelier to Kreps, Harrison and Pliska

The Mathematics

Bachelier gained his mathematics degree at the Sorbonne in 1895. He studied under an impressive lineup of professors including Paul Appell, Émile Picard, Joseph Boussinesq and Henri Poincaré.

Appell was a prodigious problem solver with little taste for developing general theories, and although he gave his name to a sequence of polynomials, his numerous contributions to analysis, geometry and mechanics are little remembered today. Picard's name, by contrast, is familiar to any undergraduate mathematician. It is attached to theorems in analysis, function theory, differential equations and analytic geometry. He also had the reputation for being an excellent teacher. In his obituary of Picard in 1943, Hadamard wrote,

> A striking feature of Picard's scientific personality was the perfection of his teaching, one of the most marvellous, if not the most marvellous [teacher], that I have ever known.

Boussinesq made contributions across mathematical physics, notably in the understanding of turbulence and the hydrodynamic boundary layer. It is from Boussinesq that Bachelier learned the theory of heat and it was on Boussinesq's work in fluid mechanics that Bachelier's second 'thesis' (really an oral examination) was based. The purpose of the second thesis was to test the breadth and teaching abilities of the candidate.[1] Bachelier's subject involved the motion of a sphere in a liquid and had the less than catchy title 'Résistance d'une masse liquide indéfinie pourvue de frottements intérieurs, régis par les formules de Navier, aux petits mouvements variés de translation d'une sphère solide, immergée dans cette masse et adhérente à la couche fluide qui la touche'.

---

[1] See Taqqu (2001). Doctoral candidates in France are subjected to the same test today.

His first degree required Bachelier to pass examinations in mechanics, differential and integral calculus and astronomy. In 1897 he also took Poincaré's examination in mathematical physics (see Taqqu 2001). Poincaré, an extremely important and influential mathematician, was professor of mathematical physics and probability at the Sorbonne during Bachelier's first degree, transferring to the chair of celestial mechanics in 1896. It was Poincaré who introduced Bachelier to probability theory, a subject which had rather fallen out of fashion in France since the great treatise of Laplace: *Théorie analytique des probabilités*, whose final edition was published in 1820.[2] Nonetheless, there was a considerable body of work to draw upon, generally couched in the language of gambling, and Bachelier had Joseph Bertrand's highly accessible book *Calcul des probabilités* from which to learn the basics.

This training in probability and the theory of heat, combined with hands-on knowledge of the stock exchange, provided Bachelier with the tools that he needed to write his remarkable thesis. Precisely this combination of mathematical ideas lays the foundation for the modern theory of Brownian motion. Bachelier is generally now credited with being the first to introduce this mathematical process, but there are earlier claims. In 1880, Thorvald Thiele, who taught Neils Bohr mathematics as professor of astronomy in Copenhagen, published (simultaneously in Danish and French) an article on time series which effectively creates a model of Brownian motion.[3]

Bachelier's Brownian motion arises as a model of the fluctuations in stock prices. He argues that the small fluctuations in price seen over a short time interval should be independent of the current value of the price; implicitly, he also assumes them to be independent of the past behaviour of the process. With these assumptions and the Central Limit

---

[2]The first edition (1812) was dedicated to 'Napoleon-le-Grand', under whom Laplace had briefly been Minister of the Interior. In his memoirs, written on St Helena, Napoleon says that he removed him from office after only six weeks 'because he brought the spirit of the infinitely small into the government'. Nonetheless, it was wise to maintain Laplace's allegiance and he was raised to the Senate. By 1814 it was evident that the Empire was falling and the dedication was replaced in the editions of 1814 and 1820 by forewords explaining how the new editions differed from the old. The political opportunist Laplace, who had helped develop the metric system for the revolutionary government, was named a Marquis in 1817, after the restoration of the Bourbons.

[3]It is sometimes argued that Thiele's great contribution to actuarial science, his differential equation for the net premium reserve as a function of time for a life insurance policy, is the first rudiment of stochastic calculus in the mathematics of insurance and finance. Although dated to 1875, it was never published. For more on Thiele see Lauritzen (2002).

Theorem he deduces that increments of the process are independent and normally distributed. In modern language, he obtains Brownian motion as the diffusion limit (that is, as a particular rescaling limit) of a random walk.

Having obtained the increments of his price process as independent Gaussian random variables, Bachelier uses the 'lack-of-memory' property for the price process to write down what we would now call the Chapman–Kolmogorov equation and from this derives (not completely rigorously) the connection with the heat equation. This 'lack-of-memory property', now known as the Markov property, was formalized by A. A. Markov in 1906 when he initiated the study of systems of random variables 'connected in a chain', processes that we now call Markov chains in his honour. Markov also wrote down the Chapman–Kolmogorov equation for chains but it was another quarter of a century before there was a rigorous treatment of Bachelier's case, in which the process has continuous paths.

The name Brownian motion derives from a very different route. In science it is given to the irregular movement of microscopic particles suspended in a liquid (in honour of the careful observations of the Scottish botanist Robert Brown, published in 1828). It was in Einstein's 'miraculous year', 1905, that he, unaware of Bachelier's work, introduced his mathematical model of Brownian motion, although he was cautious in his claims, saying that 'it is possible that the motions described here are identical with so-called Brownian molecular motion; however, the data available to me on the latter are so imprecise that I could not form a judgement on the question'. Einstein's motivation was quite different from Bachelier's. Inspired by Boltzmann's 1896 and 1898 work on the kinetic theory of matter,[4] he was looking for ways to verify the existence of atoms. Although the ultramicroscope had brought the observation of molecules closer, it remained impossible to measure their velocities. Einstein introduced the mean-square displacement of the suspended particles as the primary observable quantity in Brownian motion. He proved that under the assumptions of the molecular theory of heat, bodies of diameter of the order of 10^{-3} millimetres suspended in liquids must perform an observable random motion. His doctoral thesis already contains a derivation of the diffusion coefficient in terms of the radius of the suspended particles and the temperature and viscosity of the liquid.

[4] Just as Einstein was unaware of Bachelier, so Bachelier was unaware of Boltzmann's work, which was translated into French in 1902 (Taqqu 2001).

His first paper on Brownian motion, 'On the motion of particles suspended in fluids at rest implied by the molecular–kinetic theory of heat', contains a new derivation making use of methods of statistical physics. Einstein also derives the connection between Brownian motion and the heat equation.

He assumes that each suspended particle executes a motion that is independent of all other particles. He also requires the existence of a time interval that is short compared with the times at which observations are made and yet sufficiently long that the motions of any one suspended particle in successive time intervals can be regarded as independent. The argument for the existence of such time intervals is as follows. The physical explanation of Brownian motion is that the suspended particles are subject to haphazard collisions with molecules. By the laws of mechanics, the motion of a particle is determined not just by these impulses but also by its initial velocity. But for a particle of this size, over any 'ordinary' interval of time, the effect of the initial velocity is negligible compared with that of the impulses received during that time interval. This means that the displacement of the particle during such a time interval is approximately independent of its entire previous history. Einstein goes on to describe the motion of the suspended particles in terms of a probability distribution that determines the number of particles displaced by a given distance in each time interval. Using the assumed independence property and symmetry of the motion process he shows that this probability distribution is governed by the heat equation. Combined with his expression for the diffusion coefficient, Einstein obtains from this the probability density function and thence an expression for the mean-square displacement of a suspended particle as a function of time. He suggests that it could be used for the experimental determination of Avogadro's number. At the same time, the Polish physicist Smoluchowski published an analogous theory of Brownian motion, also focusing on the mean-square displacement of the suspended particle. His expression for this quantity differs from Einstein's only by a factor of $\frac{64}{27}$. In 1908, Langevin published a short note showing that, after correction, Smoluchowski's approach leads to Einstein's formula for the mean-square displacement. He uses his own approach to throw further light on the theory and arrives at what is now known as the Langevin equation for the position u of the suspended particle:

$$m\frac{\mathrm{d}u}{\mathrm{d}t} = -6\pi a\mu u + X,$$

83

where a is the radius of the particle, m is its mass and μ is the viscosity of the liquid in which it is suspended. He adds that (loosely translated) 'about the complementary force X we know that it is indifferently positive and negative and that its size is such that the agitation of the particle is maintained'. It was thirty-four years before Doob gave a precise mathematical meaning to Langevin's equation. In 1909, Jean Perrin completed Einstein's programme by experimentally obtaining Avogadro's constant as $N \approx 7 \times 10^{23}$, confirming estimates obtained in other ways, work which won him the 1926 Nobel Prize in physics.

Bachelier would not have been aware of the earlier work of Rayleigh, who by 1880 had already made the connection between random phase and the heat equation. Rayleigh was concerned with the superposition of n oscillations. The simplest instance of his results corresponds to coin tossing. It was Pólya who made Rayleigh's results known in Paris in 1930, by which time Bachelier was a professor in Besançon (see Taqqu 2001). Indeed the synthesis of the theories of Rayleigh and Einstein had to wait for the 1930s revival of probability theory.

Bachelier himself believed that his most important achievement was the systematic use of the concept of continuity in probabilistic modelling. He regarded continuous distributions as fundamental objects rather than just mathematical inventions for simplifying work with discrete distributions. From a modern perspective his greatest insight was the importance of *trajectories* of stochastic processes. After defending his thesis, Bachelier published an article in 1901 in which he revises the classical theory of games from what he calls a 'hyperasymptotic' point of view. Whereas the asymptotic approach of Laplace deals with the Gaussian limit, Bachelier's hyperasymptotic approach deals with trajectories and leads to what we would now call a diffusion approximation. Bachelier was well aware of the importance of his work. He wrote in 1924 that his 1912 book *Calcul des probabilités*, Volume 1 (Volume 2 was never written) was the first that surpassed the great treatise by Laplace (see Courtault et al. 2000).

Meanwhile, Einstein was quickly informed by colleagues that his predictions really did fit, at least to order of magnitude, known experimental results for Brownian motion, and later in 1905 he wrote a further paper more boldly entitled 'On the theory of Brownian motion'. Here we meet the first discussion of the fact that our mathematical theory predicts that Brownian paths are highly irregular objects. Indeed Einstein points out that his equation for mean-square displacement cannot hold for small times t since that equation implies that the mean velocity of the suspended particle over a time interval of length t tends to infinity as $t \to 0$. This stems from the assumption that the motions of the particle over successive small time intervals are independent, an approximation which breaks down for very small t. He estimates that the instantaneous velocity of the suspended particle changes magnitude and direction in periods of about 10^{-7} seconds.

Experimental results also pointed to the extremely irregular trajectories. In his 1909 article and later in his popular book *Les Atomes*, Perrin describes how the paths apparently have no tangent at any point. He says,

> C'est un cas où il est vraiment naturel de penser à ces fonctions continues sans dérivées que les mathématiciens ont imaginées et que l'on regardait à tout comme des simples curiosités mathématiques, puisque l'expérience peut les suggérer.

In other words,

> In this case it really is natural to consider the idea of continuous functions with no derivative—a construct generally regarded simply as a mathematical curiosity. Here, they are suggested by experiment.

Norbert Wiener liked to quote Perrin's observations and it was Wiener, in 1923, who finally produced a mathematically rigorous pathwise construction of Brownian motion. His approach was to construct a probability measure on the space of continuous real-valued functions on the positive half-line (in other words, on the space of continuous paths in $\mathbb{R}$) such that the increments in disjoint time intervals are Gaussian. The idea of establishing a mathematical theory of probability based on integration and measure can be traced to Borel, who proved a version of the strong law of large numbers in terms of Lebesgue measure in 1909, but in order to discuss the whole time evolution of a stochastic process, rather than just a snapshot of it at some fixed time, one needs a theory of integration and measure on function spaces and this was what Wiener was able to provide.

Wiener, a child prodigy, wrote his doctoral thesis (aged just eighteen) on mathematical logic under the supervision of Karl Schmidt. He then won a Harvard travelling scholarship and spent the period June 1913–April 1914 in Cambridge, where he went with the intention of working with Bertrand Russell.[5] On Russell's advice he took a course on analysis

---

[5] In his autobiography, Russell (1967) included the text of a letter from Wiener's father, Leo, Professor of Slavic Languages at Harvard, introducing his son. The letter lists various

from G. H. Hardy, which he describes as being a 'revelation' (see Wiener 1953). Starting from mathematical logic it covered the theory of sets, the Lebesgue integral, the general theory of functions of a real variable and the theorem of Cauchy. It even provided what Wiener regarded as an acceptable logical basis for the theory of functions of a complex variable. His reading course with Russell introduced him to the work of Einstein and while spending Christmas with his family in Munich he assiduously read various works assigned by Russell including Einstein's original papers. It was in the summer of 1919, just prior to taking up a position at MIT, that a young mathematician called Isaac Barnett (who spent most of his career at the University of Cincinnati) called on Wiener. Barnett worked in functional analysis, an area in which Wiener wanted to work, and Wiener credited him with suggesting to him the problem of integration on function space. During his first year at MIT, Wiener found a solution to this problem using some ideas worked out by P. J. Daniell, who was then working at Rice University. But he found this work to be lacking in content and so he set out to look for a physical theory that would embody a similar logical structure. He found this in Brownian motion.

Wiener's seminal paper 'Differential space' converts his previous work on integration on function space (and consequently integration in infinitely many dimensions) into a study of Brownian motion. The 'differences' are the Gaussian increments of the process. He says,

> The present paper owes its inception to a conversation which the author had with Professor Lévy in regard to the relation which the two systems of integration in infinitely many dimensions—that of Lévy and that of the author—bear to one another. For this indebtedness the author wishes to give full credit.

His paper begins with a justification of Einstein's model of Brownian motion and cites F. Soddy's translation of Perrin's 'Mouvement brownien et réalité moléculaire' ('Brownian motion and molecular reality'):

> One realizes from such examples how near the mathematicians are to the truth in refusing, by a logical instinct, to admit the pretended geometrical demonstrations, which are regarded as experimental evidence for the existence of a tangent at each point of a curve.

Here then is physical substance for his work. The non-differentiability of Brownian paths that he goes on to prove is not just a mathematical

---

professors with whom young Norbert had studied, and Russell cannot resist adding a footnote which reads 'Nevertheless he turned out well.' The Wiener family history is an interesting story in its own right: see Conway and Siegelman (2005).

curiosity, but reflects Perrin's observations of highly irregular molecular motion.

The probability measure that he constructed is now known as Wiener measure and the mathematical model of Brownian motion is often called the Wiener process. Having constructed the measure, Wiener verified that for any fixed time t the probability of differentiability in t is zero, then he establishes that the probability of satisfying a Hölder condition of order $\frac{1}{2} - \epsilon$ over any interval is one. This quantifies the 'roughness' of the path. Finally, Wiener gives the law of the Fourier coefficients. From this one can develop (on the interval $(0, 2\pi)$) a function that vanishes at zero in the form

$$X_t = \xi_0 \frac{t}{\sqrt{2\pi}} + \sum_1^\infty \frac{\xi_n(1 - \cos nt) + \xi_n' \sin nt}{n\sqrt{\pi}},$$

where $\xi_0, \xi_1, \xi_1', \ldots$ is a sequence of independent $N(0, 1)$ random variables. This is the Fourier–Wiener series implicit in the 1923 work and explicit in his 1933 collaboration with Paley and Zygmund.

Whereas Wiener constructed the continuous process that we call Brownian motion directly, Bachelier's approach provided a passage from the discrete to the continuous, but Bachelier simply did not have at his disposal the mathematical machinery to make his hyperasymptotic theory rigorous. It was Kolmogorov,[6] in the famous 1931 paper 'Über die analytischen Methoden in der Wahrscheinlichkeitsrechnung' ('On analytical methods in probability theory'), who made rigorous the passage from discrete to continuous schemes. He does this by extending Lindeberg's method for proving the Central Limit Theorem to this setting. The Kolmogorov partial differential equations can then be obtained from the difference equations that hold in discrete time. Bachelier's influence is evident: Kolmogorov credits Bachelier as being the first to make systematic the study of the case where the transition probability $P(t_0, x, t, y)$ depends continuously on time.

Kolmogorov introduced a Markov transition function as a family of stochastic kernels that satisfy the Chapman–Kolmogorov equation. He himself called it the Smoluchowski equation as Smoluchowski had already written down a special case. The central idea of the paper is the introduction of local characteristics at each time t and the construction of transition functions by solving differential equations involving

---

[6]For more on Kolmogorov's contributions to probability theory we refer to Dynkin (1989) and the other memorial articles in the same volume.

these characteristics. The 'Bachelier case' is covered in the last part of the paper, where he treats a class of real-valued transition functions for which the 'drift' and 'diffusion' coefficients can be defined. (The terminology is due to Feller.) Under additional regularity assumptions on the transition probabilities he then proves that they satisfy the Fokker–Planck equation.[7] Kolmogorov's work had a powerful effect on the development of probability theory. Now the search was on for distribution functions of continuous limiting processes without recourse to passage to the limit from an approximating sequence.

Kolmogorov's highly influential 1933 monograph *Grundbegriffe der Wahrscheinlichkeitsrechnung* ('Elements of probability theory') transformed the nature of the calculus of probabilities. Although it had already been understood that the basic manipulations of probability were the same as those of measure theory, the relationship between the two was not sufficiently well formulated to be useful. The principal problem was the definition of conditional probabilities. Bachelier manipulated these in a non-rigorous way with ease. Kolmogorov gave them a firm mathematical foundation by defining conditional probabilities and expectations as random variables whose existence is guaranteed by the Radon–Nikodým Theorem. Finally, 'calculus of probabilities' had become a respectable part of mathematics: 'probability theory'.

The results in Kolmogorov's monograph led to a measure on the space of all functions for which the finite-dimensional distributions are specified, but whereas Wiener measure was defined on continuous functions, so that Wiener's Brownian motion has continuous trajectories, here there is no regularity. The question of how to restrict the measure to classes of functions with good regularity properties remained. A first step in this direction was Kolmogorov's continuity criterion, first published in a joint paper with Aleksandrov in 1936.

The next big milestone in the development of measure theory on function spaces was the publication of Doob's 1953 monograph, but before charting Doob's contributions we turn to his friend and colleague William Feller. Feller and Doob's books, written at the same time, were of key importance in shaping the future of probability theory. Feller's two-volume work remains a standard reference over half a century after it was first published. It was meant to be three volumes, but sadly Feller died before the third, to be devoted to stochastic processes, could be

---

[7]At the time of writing the article, Kolmogorov was unaware of earlier work of Fokker and Planck in which they had written down this equation in a special form, but after 1934 he referred to it as the Fokker–Planck equation.

written (see Doob 1970). Doob and Feller also cast in stone a large part of the terminology of modern stochastic analysis. Inspired by his review of Ville's book, in 1940 Doob began his study of martingales, calling them 'processes with property E', but in his book he reverted to Ville's terminology and he later said that he attributed much of their success to the name. This was not his only concession. He had wanted to use the term 'chance variable' but it seems that after an argument with Feller, who preferred the term 'random variable', they flipped a coin and Feller won (see Snell 2005). The term random variable is used in both books and is firmly established as the norm.

Feller's mathematical legacy is enormous. In 1936 he extended the work of Kolmogorov, discussing existence and uniqueness for general Markov processes with jumps. Generalizing Kolmogorov's parabolic differential equations, he provided a powerful analytic tool with which to study transition probabilities. The integrodifferential equations were his main interest rather than sample path properties of the process, but the two turned out to be intimately related. Feller applied semigroup theory in this setting and observed that the appropriate boundary conditions for the Kolmogorov equation not only specified the domain of the infinitesimal generator of the corresponding semigroup, but also characterized the behaviour of the trajectories of the process at the boundaries of the state space. Although the definitive result is only found in the 1965 book of Itô and McKean, this work was essential to the beautiful characterization of general one-dimensional diffusions in terms of a scale function and a speed measure. Any one-dimensional diffusion can be obtained from a Brownian motion by first time-changing that Brownian motion (in a way determined by the speed measure) and then applying a strictly increasing function (the inverse of the scale function) to the time-changed process.

Even on the purely mathematical side Feller did much more. For example, he was the main architect of renewal theory. But as a permanent visiting professor at the Rockefeller University he also developed close collaborations with population geneticists and it was Feller's beautiful paper of 1951 that initiated a cross-fertilization of ideas between population genetics and diffusion theory, which is of enormous importance to the present day.

At this point in our history it is still the case that paths played little mathematical role. For Kolmogorov and Feller alike, the integrodifferential equations that governed the transition probabilities of the processes were the main object of study. But for Bachelier things were different.

Although he did not have the mathematical technique at his disposal to rigorously construct Brownian motion as a continuous stochastic process, and many of his results are calculated from asymptotic Bernoulli trial probabilities, it is clear from his thesis and later work that Bachelier did think in terms of trajectories. The most striking example in the thesis, and the one that evidently impressed Poincaré, was the elegant argument that he gives for the reflection principle. The reflection principle is generally attributed to Desiré André, who proved it in the purely combinatorial form, as credited by Bachelier. André was a student of Joseph Bertrand and one can actually find the reflection principle in the context of gambling losses in Bertrand's 1888 book *Calcul des probabilités*.

Bachelier's argument justifying the reflection principle is not rigorous because it requires the strong Markov property of Brownian motion. This property was only properly formulated by Doob in the 1940s and was finally established for Brownian motion by Hunt in 1956. Symmetry arguments similar to that proposed by Bachelier for the reflection principle were also used to great effect by Paul Lévy, who established many detailed results about the paths of Brownian motion. In the words of Loève in his obituary of Lévy, 'But above all he is a traveller along paths (this is why for him the Markov property is always the strong Markov property).' And as Doob remarked (see Doob 1970),

> Lévy was not a formalist and in particular was not sympathetic to the delicate formalism that discriminates between the Markov and strong Markov properties.

At the same time as Paley, Wiener and Zygmund made explicit the Fourier–Wiener series, Paul Lévy had been thinking along similar lines. Indeed, Brownian motion is already discussed in his 1937 book *Théorie de l'addition des variables aléatoires*. Lévy observed that one can replace the (trigonometric) Fourier basis functions with other choices to obtain other series expansions of the Wiener process:

$$X_t(\omega) = \sum a_n(t)\xi_n(\omega).$$

In an elegant paper of 1961, Ciesielski chooses the Haar basis. Then the $a_n(t)$ are triangular functions supported by the dyadic intervals and the study of the series is quite simple. Lévy wrote,

> At the beginning of 1934, I suddenly noticed that any stable law leads, as does the Gaussian, to a random function that we can obtain, like that of Wiener, by an interpolation method. I then decided to find the

general form of a function $X(t)$ with independent increments, in other words of an additive process.

Taken from Loève (1973)

The interpolation consists of determining the probability distribution of $X(t + \frac{1}{2}h)$ when $X(t)$ and $X(t + h)$ are known. This led him to the definition of what are known as 'Lévy processes', another class of mathematical models now widely used in financial applications.

Just before World War II Bachelier published a new book on Brownian motion and in 1941 a further paper. In his 1970 memoirs Lévy says that he only became aware of this work after the war. Some of the results there are rederived in Lévy's own book on Brownian motion published in 1948.[8] We do not reproduce here the well-documented tale of Lévy's role in Bachelier's difficult career.[9] As is evident from his thesis, Bachelier could be sloppy. Although his results are correct, many of his assumptions are implicit rather than explicit and this can make his work hard to follow. Lévy appears to have been genuinely misled. Bachelier defines Brownian motion to be a limit as $\tau \to 0$ of random functions that are linear on each interval $[n\tau, (n + 1)\tau]$ with derivatives v or $-v$ taken at random with equal probabilities. The dependence of v on τ, in fact $v = c\tau^{1/2}$, is not made explicit and without this rescaling the sequence of processes evidently does not converge. Lévy interpreted this as a gross error and read no further. The two men were eventually reconciled and Lévy conceded priority over a number of results to Bachelier.

Fundamental to Bachelier's calculations is what he calls the *true* price of a security. All his calculations are in terms of the true price for which 'the mathematical expectation of the speculator is null'. Combined with the implicit lack-of-memory property of the price process, Bachelier is saying that the true price is a *martingale*. It was also Lévy who, in 1934, formalized the concept of a martingale (although the name was coined by Ville in 1939). He introduced the concept in an attempt to preserve the law of large numbers while relaxing independence assumptions (similar considerations motivate Markov dependence). Guided by results obtained for sums of independent random variables he initiated the study of martingales, but it was Doob who transformed the subject and through him martingales became a powerful tool in both probability and analysis.

---

[8] Being Jewish, Lévy was not allowed to publish during the war, some of which he spent in hiding.

[9] See Courtault and Kabanov (2002) for an account of this affair.

Joseph Doob came to probability theory from complex analysis. His 1932 doctoral thesis was entitled 'Boundary values of analytic functions'. He saw the connection between martingales and harmonic functions (also published in two short notes by Kakutani in 1944/45) and based on this he worked to develop a probabilistic potential theory. Martingale theory is the focus of one of the chapters (nearly one hundred pages long) of his 1953 book *Stochastic Processes*, one of the most influential books on probability theory ever written. Doob's work in the area was heavily influenced by Bachelier. In October 2003 he wrote,[10]

> I started studying probability in 1934, and found references to Bachelier in French texts, along with references to the reflection principle of Desiré André which I think I remember Bachelier used heavily. I looked up both Bachelier and André and learned a lot. Later I learned that Bachelier's work was rediscovered later by Lévy and others. Of course the rigorous proofs of Bachelier's results for Brownian motion had to wait for rigorous mathematical definitions of Brownian motion and the development of suitable techniques. As I remember Bachelier's writing, he scorned other writers and asserted that only he had obtained new results. In his day, and also considerably later, probability was not a respectable part of mathematics and one man's probability theory was another man's nonsense. The ideas of Bachelier and André made a permanent impression on me, and influenced my work on gambling systems and later on martingale theory.

Doob was also one of the first people to study stochastic differential equations. In the introduction to a paper of 1942 he says,

> A stochastic differential equation will be introduced in a rigorous way to give a precise meaning to the Langevin differential equation for the velocity function $dx(s)/ds$.

But the central figure in the development of stochastic differential equations was the Japanese mathematician Kiyosi Itô. Itô drew on all the great figures in our story so far. Doob had used measure theory to make sense of the intuitive ideas of sample paths. Kolmogorov and Feller had instead emphasized the connections between Markov processes and partial differential equations. Itô's first paper on stochastic integration, written in 1944, is extremely short. There is very little preamble, he refers to Lévy's book and to Doob's 1937 paper, 'Stochastic processes depending on a continuous parameter', for the definition of Brownian motion and to Paley and Wiener's work of 1934 for the special case of the stochastic

---

[10]Personal communication to the authors.

integral where the integrand is deterministic. However, in his collected works, Itô explains his motivation:

> In these papers I saw a powerful analytic method to study the transition probabilities of the process, namely Kolmogorov's parabolic equation and its extension by Feller. But I wanted to study the paths of Markov processes in the same way as Lévy observed differential processes.

Thinking about the behaviour of a Markovian particle over infinitesimal time increments, Itô formulated the notion of stochastic differential equations governing the paths of Markov processes. If W is a standard Wiener process, then Itô's equation for the position of a particle following a Markov process could be written as

$$dX_t = \mu(X_t)\,dt + \sigma(X_t)\,dW_t.$$

In his first paper on stochastic integration in 1944 he made sense of the notion of solution to this equation, that is, he gave a rigorous mathematical meaning to the stochastic integral. He also states what one might call 'the fundamental theorem of calculus' for functions of Brownian motion. In his second paper of 1951 he stated and proved what is now known as Itô's formula, which makes the connection with Kolmogorov's partial differential equations. Whereas the Wiener integral is the integral of a deterministic function against white noise, time playing no role, time is crucial in the Itô integral and, moreover, the integrand can itself be a random function.

In Doob's 1953 book, Itô's stochastic calculus is extended to processes with orthogonal increments and then to processes with conditionally orthogonal increments, that is martingales. However, Doob had to make an assumption. In order to be able to define the stochastic integral with respect to the martingale M, he required the existence of a non-random increasing function $F(t)$ such that $M_t^2 - F(t)$ is also a martingale. (When M is Brownian motion, the function $F(t) = t$ has this property.) For discrete-parameter martingales, the analogous property follows from the Doob Decomposition Theorem, which allows one to write a submartingale (uniquely) as the sum of a martingale and a process with increasing paths, starting from zero, and with the property that the process at time n is measurable with respect to the sigma-field generated by information available up to time $n - 1$. The continuous-time version of this result, under a certain uniform integrability assumption, is due to Meyer (1962). Uniqueness of what is now called the *Doob–Meyer decomposition* came a year later in Meyer (1963). In his first paper Meyer

93

proposes, as an application of the decomposition theorem, an extension of Doob's stochastic integral. A systematic development of these ideas is provided by Courrège (1963), but it was left to Kunita and Watanabe (1967) to provide the analogue of Itô's formula for these more general stochastic integrals.[11]

Up to this point, stochastic integration was intimately bound with the theory of Markov processes. This came about through a measure-theoretic constraint: the underlying filtration of σ-algebras was assumed to be quasi-left continuous. (This means that the process has no *fixed* points of discontinuity.) In 1970, Doléans-Dade and Meyer removed this hypothesis and stochastic integration became purely a martingale theory (or, more precisely, a semimartingale theory). From the mathematical finance point of view, this was a key ingredient in the fundamental papers of Harrison and Kreps (1979) and Harrison and Pliska (1981).

We should not close this section without mentioning the extraordinary story of Vincent Doeblin, without whose untimely death in World War II the whole story could have unfolded differently.[12] Born Wolfgang, he was the second son of Alfred Döblin, author of *Berlin Alexanderplatz*, who being Jewish and left-wing saw his books publicly burned by the Nazis before fleeing Germany in 1933 after the burning of the Reichstag. He fled first to Zurich and then to Paris, where Wolfgang enrolled at the Sorbonne with the intention of eventually studying statistics in relation to political economics; but having gained his certificates in general mathematics, rational mechanics and probability theory he turned to the theoretical probability in which Paris and Moscow were leading the world. At the end of 1935 Wolfgang began research into the theory of Markov chains under the supervision of Maurice Fréchet and Paul Lévy, with whom he wrote his first paper (Doeblin and Lévy 1936). In 1936 he took French nationality and changed his name to Vincent. Two years later, on completion of his doctorate, he signed up for two years' military service. In September 1939 as a telegraphist in the newly formed 291st infantry regiment he was sent to the Ardennes. He told his fellow soldiers that he was from Alsace. Although initially he abandoned scientific work, a letter from Fréchet with an invitation to collaborate seems to have inspired him to resume his mathematical research under what must have been extraordinarily difficult conditions. He wrote to Fréchet

---

[11]See Jarrow and Protter (2004) for a more thorough account of these developments.

[12]For a complete account see Bru and Yor (2002).

that he was writing the developed proofs of his note on Kolmogorov's equation, which had been published in 1938, shortly before his departure for military service. However, on the night of 20–21 June 1940, Doeblin found himself with the remains of his regiment in the Vosges, completely surrounded by German soldiers. After walking all night, he reached the village of Housseras, which had been taken by the Germans. He burned all his papers and shot himself in the head.

It was long believed that the work on Kolmogorov's equation to which Doeblin had been devoting himself had been lost. But in fact earlier in 1940, Doeblin had sent a sealed envelope, a *pli cacheté*, from the front line to the Académie des Sciences in Paris. The deposit of a *pli cacheté* allowed an author to establish priority over a scientific result when temporarily unable to publish it in its entirety. Lévy resorted to a *pli* when racist laws prevented him from publishing. Doeblin wrote to Fréchet announcing the dispatch of his *pli* about Kolmogorov's equation, but after the war the *pli* was forgotten and it was not until Doeblin's letter of 12 March 1940 was discovered in the archives of Maurice Fréchet that its existence came to light. In the year 2000, at the request of Wolfgang's brother, Claude Doblin, the *pli* was finally opened.

Doeblin had already made fundamental contributions to the theory of Markov processes. In Volume 2 of his great treatise, Feller describes Doeblin's work on sums of independent random variables as 'masterful'. Lévy dedicated an article to Doeblin in 1955. There he wrote,

> on peut compter sur les doigts d'une seule main les mathématiciens qui, depuis Abel et Galois, sont morts si jeunes en laissant une œuvre aussi importante...

> one can count on the fingers of one hand those mathematicians who, since Abel and Galois, died so young while leaving such an important oeuvre...

This takes account only of Doeblin's published work. In May 2000, Doeblin's *pli cacheté* was opened. The title of the *pli* is 'Sur l'équation de Kolmogoroff' ('On Kolmogorov's equation'). It is concerned with the construction and study of continuous Markov processes, especially one-dimensional diffusions. Although it predates the great development of the theory of martingales, it analyses the paths of an inhomogeneous real-valued diffusion from a martingale perspective. Using quadratic variation as a 'clock' he uses time transformations to obtain a mathematically rigorous solution for the 'change-of-variable problem' for a class of one-dimensional diffusions. This is a very close relative of the

95

Itô formula. Had this extraordinary document been opened sooner, the history of stochastic processes would have looked very different.

THE ECONOMICS

As we have seen, Bachelier and his work appear as a continuous thread through the development of twentieth century stochastic analysis. He published his thesis and a book on probability theory before World War I and was personally known to leading figures in the French probability community throughout his career. His work was cited in two of the century's most influential works on probability, the Kolmogorov (1931) paper on analytic methods in probability and Doob's *Stochastic Processes*. In short, no one could say they had not been told.

The situation in the worlds of economics and finance could not have been more different. Although the 1908 book of de Montessus on probability and its applications devotes a whole chapter to finance that is based on Bachelier's thesis (see Taqqu 2001), his work made almost no impact when it appeared and lay forgotten for more than fifty years before it finally arrived on the desks of financial economists. We have already described the way in which this happened: in the mid 1950s statistician Jimmy Savage sent postcards to his economist friends alerting them to Bachelier's work. The postcard addressed to Paul Samuelson arrived at an opportune moment, because Samuelson was at the time very much concerned with questions of option and warrant valuation and had at least one PhD student, Richard Kruizenga, working in the area. The Bachelier thesis was duly acquired, and Samuelson commissioned the translation by A. James Boness that now appears, together with work on options and price modelling by Kruizenga and by Benoît Mandelbrot and others, in Paul Cootner's 1964 book *The Random Character of Stock Market Prices*. Asked what he learned from Bachelier's thesis, Samuelson said 'it was the tools'—the panoply of mathematical techniques deployed by Bachelier encompassing Brownian motion, martingales, Markov processes, the heat equation. These were just the things needed for Samuelson's own programme.

Would history have been different if economists had been alerted to Bachelier a few decades earlier? We do not believe so. For one thing, there was no great interest in the subject of option valuation. As Cox, Ross and Rubinstein put it in their 1979 paper, 'Options have been traded for centuries, but remained relatively obscure financial instruments until the introduction of a listed options exchange in 1973.' In fact, 'optionality' in various guises is a ubiquitous feature of financial markets, and how to value it is a key component of asset valuation in general, but this point was not widely appreciated before the massive expansion of financial markets activity in the latter third of the twentieth century. This expansion, in turn, could not have occurred without the contemporary developments in computer technology. In earlier days there was no way to hedge an option contract: markets were too illiquid, costs too high and information too scanty. Effective management of option risks depends on having a 'deep' (implying large) market and trading on a sufficiently fast timescale. Before the modern era of massive computational power and cheap memory, none of this was feasible. Computer technology is the third leg—alongside the economics and the mathematics—on which modern financial markets stand.

An obvious deficiency of Bachelier's Brownian motion model of asset prices is that the price at any one time, being normally distributed, can be negative.[13] To remedy this, Samuelson introduced the *geometric Brownian motion* model in which the asset price $S(t)$ is given by

$$S(t) = S(0)\exp(at + \sigma W(t)), \tag{3.1}$$

where $W(t)$ is Brownian motion and a, σ are constants. Since $W(t) \sim N(0, t)$ we have $\mathbb{E}[e^{\sigma W(t)}] = \exp(\frac{1}{2}\sigma^2 t)$, so if we take

$$a = \alpha - \tfrac{1}{2}\sigma^2, \tag{3.2}$$

then $\mathbb{E}[S(t)] = S(0)e^{\alpha t}$. Thus α is the expected growth rate. The parameter σ is known as the *volatility* and measures the standard deviation of log-returns[14]: the standard deviation of $\log(S(t + h)/S(t))$ is $\sigma\sqrt{h}$.

The move from arithmetic to geometric Brownian motion was, on one level, a simple expedient to secure positive prices, but on another level it had a profound influence on the whole mathematical development of the subject. The exponential is a *non-linear* function, and analysis of non-linear transformations of Brownian motion necessarily involves Itô calculus. Applying the Itô formula to $S(t)$ given by (3.1), (3.2) shows that $S(t)$ satisfies the stochastic differential equation (SDE)

$$dS(t) = \alpha S(t)\,dt + \sigma S(t)\,dW(t). \tag{3.3}$$

---

[13]Bachelier realizes this, but assumes that it happens with an effectively negligible probability (see p. 29).

[14]For small h, $\log(S(t + h)/S(t)) \sim (S(t + h) - S(t))/S(t)$, the 'return' obtained by buying the asset at time t and selling at time $t + h$.

This is a much more intuitive representation than (3.1): we can immediately see, for example, that the average growth rate is α and that $S(t)$ is a martingale if $\alpha = 0$. And if we want to examine trading strategies for buying and selling the asset then the SDE representation is essential.

As we saw above, by 1960 the Itô integral, SDEs and the connection with the heat equation were all well understood. But we can hardly say they were well understood by the man or woman in the street, or even in the scientific laboratory. Ordinary 'Newton–Leibnitz' calculus is, and has for centuries been, used in sophisticated ways by a huge range of scientific and engineering practitioners, only a small minority of whom would describe themselves as professional mathematicians. In 1960, by contrast, stochastic calculus was a branch of pure mathematics, studied and understood by at most a few hundred specialists around the globe. There were no textbooks taking anything like an applied point of view. This situation did not, however, last for long, as study of random phenomena became increasingly important in several different applied areas. In engineering, the accent in the space programme era was on dynamical systems, and a major boost to the study of differential equations with random inputs was provided by Rudolf E. Kalman with his introduction of the Kalman filter in 1960. In fact, there is another connection with our existing cast of characters here, in that both Kolmogorov and (independently) Norbert Wiener had studied linear filtering and prediction problems in connection with military applications in World War II. Wiener's work was originally produced as a yellow-covered classified report that was dubbed the 'yellow peril' because it was so impossibly hard to understand; it eventually saw the light of day in 1949. Both authors formulated the problem in the context of stationary processes and gave answers in the form of a 'filter' in the electronics sense of the term. Revisiting the problem in the space era, Kalman adopted a 'time-domain' approach, modelling the signal to be estimated as a discrete-time dynamical system and thereby eliminating the need for stationarity. This approach met with immense success and continues in widespread use today. The continuous-time version was introduced by Bucy and Kalman (1961); now the signal is indeed a noise-perturbed differential equation. The full generality of Itô calculus is not, however, required, since all equations are linear and only the Wiener integral (the Itô integral with deterministic integrand) is needed. Attempts were soon made to extend filtering theory to non-linear systems, a subject that certainly does require the machinery of stochastic calculus. It took a few years, but by 1968 a correct formulation had been given by Bucy, Kushner,

Shiryaev and others in a series of works, which even in hindsight look impressively sophisticated.

Another natural question from the engineering standpoint is this. If we formally divide (3.3) by dt we obtain a 'differential equation',

$$\frac{\mathrm{d}}{\mathrm{d}t}S(t) = \alpha S(t) + \sigma S(t)\frac{\mathrm{d}W}{\mathrm{d}t},$$

which looks like the simple differential equation $\dot{S} = \alpha S$ perturbed by 'white noise' $\sigma S\dot{W}$. This interpretation does not make sense as it stands ($W(t)$ is not differentiable), but one could ask whether $S(t)$ is some kind of limit of a noise-perturbed dynamical system as the noise becomes 'whiter'. The answer, uncovered by Wong and Zakai (1965), was that the limit is indeed an SDE, but with the drift $\alpha S\,\mathrm{d}t$ adjusted by the famous *Wong–Zakai correction term*.

Like filtering theory, this result, valuable in itself, was instrumental in bringing stochastic calculus to a wider audience. For a brief period there were actually two competing stochastic integrals in the field. R. L. Stratonovich, studying problems in telecommunications, gave a different definition under which, unlike the Itô integral, the ordinary rules of calculus were satisfied. It was quickly realized, however, that the two integrals are related to each other in a simply stated way and that Itô's is the more fundamental of the two.[15] But, for some applications, particularly in relation to stochastic differential geometry, the Stratonovich integral is the right way to go. In fact, the Wong–Zakai correction term alluded to above just expresses the difference between an Itô integral and the corresponding Stratonovich integral.

The net effect of all these developments was that by the late 1960s stochastic calculus was in far better shape from the point of view of end-users and several excellent books such as McKean (1969) or, taking a more applied view, Kushner (1967) were available to explain it all in a palatable way.

In economics, once again Samuelson had the inside track in that Henry McKean was just around the corner in the mathematics department at MIT. McKean was in fact working directly with Itô, a collaboration that led to their 1965 book *Diffusion Processes and Their Sample Paths* and, indirectly, to McKean's later solo production *Stochastic Integrals*. These remain two of the most admired books in the subject. Also in 1965, Samuelson and McKean collaborated over a paper on option pricing

---

[15]It is defined for a wider class of integrands.

(Samuelson 1965) about which we will have more to say below. A further recruit to the project was Robert Merton, who arrived as a graduate student at MIT in 1967 with a background in applied mathematics. Merton was probably the first person to appreciate clearly the connection between Itô integrals and trading strategies, and he quickly produced a series of classic papers (collected in his book *Continuous-Time Finance*) in which significant problems of financial economics were solved by methods of Itô calculus.

En Route to Black–Scholes

We do not propose to record in detail the many tortuous steps taken by researchers in search of an option pricing formula prior to the decisive breakthrough by Black and Scholes in 1973. A good impression of the various theories as of the mid 1960s can be gleaned from Cootner's book *The Random Character of Stock Market Prices*. As Harrison and Pliska put it in 1981, 'these theories, developed between 1950 and 1970, all contained *ad hoc* elements, and they left even their creators feeling vaguely dissatisfied'. Our economist friends will probably dislike this analogy, but it seems that the process the financial economists were going through bears some similarity to what was happening across the fence in stochastic analysis at very much the same time. As we have already described, the development of stochastic process theory in the 1960s was closely bound up with Markov processes. One can see this in the original 1966 edition of Meyer's *Probability and Potentials* and in the famous Kunita and Watanabe paper of 1967. As Meyer's *théorie générale des processus* gathered momentum, the relationship with Markov processes gradually withered away, leaving as a final product a *théorie des semimartingales* in which Markov processes played no direct role at all.

In a similar manner, the job of the economists in relation to the option pricing problem was to get rid of the economics. It was originally thought, reasonably enough, that since an option is a risky contract its value must have something to do both with other risky assets in the market and with investors' risk preferences. Option valuers were therefore looking at utility functions, economic equilibrium, the Capital Asset Pricing Model and so on. In the context of 'complete markets'—the Black-Scholes world—all of these are simply irrelevant. The only 'economics' left is the statement that people prefer more to less and a closely related principle, sometimes dignified with the name 'the law of one price', which, more formally stated, says that two contracts that deliver exactly the same (fixed or random) cash flows in the future must have the same value today. Otherwise one could sell the dearer one, buy the cheaper one, pocket the difference and walk away: an arbitrage opportunity. Of course, a big assumption is buried in this argument, namely 'frictionless markets', i.e. the ability to trade long and short in arbitrary amounts with no transaction costs.

Economists will say, correctly, that complete markets constitute a wafer-thin slice of economic activity and it is only on this slice that economic theory is reduced to more-better-than-less. Nonetheless, the impact of the complete markets theory has been overwhelming in terms of the development of capital markets because of its ability to make unambiguous quantitative statements.

Samuelson's 1965 paper 'Rational theory of warrant pricing' is, in the author's words, 'a compact report on desultory researches stretching back over more than a decade'. It is one of those papers that contain ad hoc elements, but aside from that it is notable for three things: the sheer quality of the intuition, the first formal analysis of American options, and the collaboration with McKean, who contributed a lengthy appendix on the American option problem. A *warrant* is an offer by a company to sell shares to investors at a stated price. From our point of view it is the same thing as a call option—there are contractual differences, but they need not detain us. Warrants can generally be exercised at any time up to some fixed expiry time T and are thus American options. There may be no expiry time, so that $T = \infty$, in which case we have a 'perpetual' warrant. The empirical facts are that perpetual warrants trade for less than the underlying stock and are exercised over time, depending on movements of the underlying price. The challenge is to determine a 'rational' price and the optimal exercise strategy. Samuelson adopted the geometric Brownian motion model (3.1), so that in particular the expected growth rate is α, i.e. $\mathbb{E}[S(t)] = S(0)e^{\alpha t}$. The 'ad hoc assumption' is that the warrant price has expected growth rate equal to a constant β *up to the time when it is optimal to exercise the warrant.* Samuelson argues that if $\beta = \alpha$ then it is never optimal to exercise before the expiry date T, so a viable theory requires $\beta > \alpha$, which in any case is necessary to compensate the warrant holder for not receiving dividends before exercise. McKean gives a precise formulation and a complete solution in terms of a free-boundary problem in partial differential equations (PDEs). In the perpetual case the PDE becomes an ordinary differential equation and the problem can be explicitly solved (otherwise numerical solution is required). The point about all of this from a

latter-day perspective is that the solution coincides with Black–Scholes if
one interprets β as the riskless rate of interest r and if $\alpha = r - q$, where
q is the dividend yield of the stock. In particular the connection with
free-boundary problems and the mathematical results slot right into the
later theory. It took a surprisingly long time, however, for a rigorous
proof to appear showing that the solution of the free-boundary problem
provides a unique arbitrage-free price. Some of the background to this
is given below.

Robert Merton's early work was not directly in option pricing but,
rather, in portfolio selection and optimal investment problems. The
techniques he introduced proved, however, to be of key importance
for the option pricing problem later on. The traditional 'mean-variance'
approach to portfolio selection, introduced by Markowitz (1952), is a
single-period theory. One starts with an endowment, which is invested
in certain securities characterized by the means and covariances of their
random returns over a fixed investment horizon. The objective is to
determine the asset allocation that achieves the maximum mean return
subject to a given constraint on the variance. The result is the well-known
'efficient frontier', which determines the optimal allocation depending
on the investor's risk appetite. This is 'one-shot' investment and does not
say how an investor should behave *over time*. Merton, in classic papers,
moved to a continuous-time setting, modelling the risky assets by lognormal diffusions as in (3.3) above, and assuming the availability of a
riskless savings account paying interest rate r. Suppose to start with
that there is just one risky asset, with price $S(t)$, and that the investor
starts with an endowment x and receives no further income. Suppose
that, at time t, the investor's portfolio value is $X(t)$, and that he invests
a fraction $u(t)$ of this value in the risky asset and consumes at rate $c(t)$.
Then the number of shares of the risky asset held is $u(t)X(t)/S(t)$. The
remaining wealth is $(1-u(t))X(t)$ and this is held in the savings account.
The increment in the investor's wealth over a short time $\mathrm{d}t$ is therefore

$$\mathrm{d}X(t) = \frac{u(t)X(t)}{S(t)} \mathrm{d}S(t) + (1 - u(t))X(t)r\,\mathrm{d}t - c(t)\,\mathrm{d}t$$
$$= \{X(t)(r + u(t)(\alpha - r)) - c(t)\}\,\mathrm{d}t + u(t)X(t)\sigma\,\mathrm{d}W(t). \quad (3.4)$$

In the first line, the three terms on the right-hand side are, respectively,
the capital gain due to movement in the asset price, the interest on the
bank deposit, and the consumption, over the interval $[t, t + \mathrm{d}t)$. If we
allow the process $u(t)$ to be a general adapted process (with some technical conditions), then (3.4) is a stochastic differential equation whose

solution $X(t)$ is the 'wealth' or 'portfolio value' process resulting from a given investment–consumption strategy (u, c). The strategy is automatically self-financing. The problem now is to choose (u, c) to maximize some measure of 'satisfaction'. Merton chose the lifetime utility of consumption criterion

$$J(x; u, c) = \mathbb{E}\left\{ \int_0^\infty e^{-\delta t} U(c(t))\, dt \right\},$$

where δ is a subjective discount factor and U is a concave 'utility' function; for example, $U(c) = c^\gamma / \gamma$ for some risk-aversion coefficient $\gamma < 1$, $\gamma \neq 0$. Of course, we have to require that the investor remains solvent, i.e. $X(t) \geq 0$ at all times; otherwise the rational investor would just rack up infinite debt. Merton uncovered the fact that this problem has an elegant and very intuitive solution. For the power utility function given above the optimal strategies are

$$u(t) = u^* = \frac{1}{1 - \gamma} \frac{\alpha - r}{\sigma^2}, \qquad c(t) = \text{const.} \times X(t).$$

The investor keeps *a fixed fraction u^* of wealth in stock*, and *consumes at a rate proportional to current wealth*. Furthermore, u^* behaves as one would expect; for example, the higher the mean growth rate α of the stock, the greater the fraction of wealth invested in it. Actually, determining this optimal strategy is a problem in *stochastic control* and it remains true to this day that the 'Merton problem' is one of a very, very small number of non-linear stochastic control problems that can be explicitly solved.

The solution of this problem was of course valuable in itself but, from the point of view of further developments in option pricing, the technique deployed to solve it was equally valuable. It involves Itô stochastic calculus in an essential way, together with a clearly formulated concept of self-financing strategies. Both of these are necessary ingredients of a Black–Scholes style theory.

THE BINOMIAL MODEL

To describe the achievements of Black, Scholes and their successors we introduce the reader at this point to the single-period binomial model (originally introduced by Cox, Ross and Rubinstein in 1979). Although at first sight extremely artificial, this model has the big advantage that the whole theory can be described in a couple of pages and the only calculation required is solution of one pair of simultaneous linear equations.

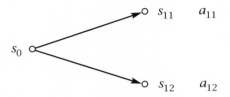

Figure 3.1. Single-period binomial tree.

The model is shown in Figure 3.1. At time 0, an asset has price S_0 equal to some value s_0. At time 1, its price S_1 is one of two known values s_{11}, s_{12} (we take $s_{11} > s_{12}$). Each of the two values occurs with strictly positive probability, but we *do not* specify what this probability is. The other asset in the market is a riskless bank account paying interest at per-period rate r, so that \$1 deposited at time 0 pays \$$R$ at time 1, where $R = 1 + r$. A contingent claim is written on the asset. This is a contract which is exercised at time 1 and has exercise value $A_1 = \$a_{11}$ if $S_1 = s_{11}$ and $A_1 = \$a_{12}$ if $S_1 = s_{12}$. It could be a call option, so that $A_1 = \max(S_1 - K, 0)$ for some strike price K, but it is not important how the values a_{11}, a_{12} are arrived at; they are completely arbitrary. We assume a frictionless market, meaning that the two assets can be traded in arbitrary amounts, positive and negative, with no costs.

There is arbitrage in this model if $Rs_0 \leqslant s_{12}$ or $Rs_0 \geqslant s_{11}$: in these cases borrowing from the bank and investing in stock, or vice versa, realizes a riskless profit with positive probability. We therefore suppose that $s_{12} < Rs_0 < s_{11}$.

Suppose that at time 0 we form a portfolio consisting of \$$B$ in the bank and N shares of stock. The value of this portfolio is $B + NS_0$ and its value at time 1 will be either $RB + Ns_{11}$ or $RB + Ns_{12}$. Suppose we choose B, N to satisfy the linear equations

$$RB + Ns_{11} = a_{11},$$
$$RB + Ns_{12} = a_{12},$$

to which the solution is

$$N = \frac{a_{11} - a_{12}}{s_{11} - s_{12}}, \qquad B = \frac{a_{12}s_{11} - a_{11}s_{12}}{R(s_{11} - s_{12})}. \tag{3.5}$$

With this choice of N, B the value of our portfolio coincides with the option exercise value, whichever way the price moves. We say the portfolio *replicates* the option payoff. By the law of one price, the value of the option at time 0 must be equal to the value of the portfolio at time 0, which is

$$A_0 \equiv B + NS_0 = \frac{1}{R}\frac{a_{12}s_{11} - a_{11}s_{12}}{s_{11} - s_{12}} + \frac{a_{11} - a_{12}}{s_{11} - s_{12}}s_0. \tag{3.6}$$

We have shown that there is a unique arbitrage-free price for the option, obtained by calculating the 'perfect hedging' strategy (B, N). This is the essence of the Black–Scholes argument. However, more can be said. We can rearrange the price formula (3.6) to read

$$A_0 = \frac{1}{R}(qa_{11} + (1-q)a_{12}),$$

where $q = (Rs_0 - s_{12})/(s_{11} - s_{12})$. Note that q does not depend on the option contract and that our no-arbitrage assumption $s_{12} < Rs_0 < s_{11}$ is equivalent to the statement that $0 < q < 1$. We can therefore interpret $q, (1-q)$ as probabilities of an upward and downward move, respectively, and rewrite (3.6) again as

$$A_0 = \mathbb{E}_q\left[\frac{1}{R}A_1\right]. \tag{3.7}$$

This states that the option value is the expected value, under the probability measure defined by q, of the discounted payoff. This probability measure is called the *risk-neutral* or *equivalent martingale* measure. The latter terminology arises from a further characterization of q, evident from (3.7). Indeed, if we put $a_{11} = s_{11}$, $a_{12} = s_{12}$, then $A_0 = S_0$ and we see from (3.7) that the process $\tilde{S}_k$, $k = 0, 1$, of discounted asset prices, defined by $\tilde{S}_0 = S_0$, $\tilde{S}_1 = S_1/R$, is a martingale. q is the unique upward probability such that this is so.

The key thing to realize here is that q is not the actual probability of an upward move. We were very careful not to state what this probability is, but only to say that upward and downward moves both occur with positive probability (and that there is no other sort of move). In technical terms, this means that the measure defined by q is 'equivalent' to the actual probability.

To summarize, we have shown that

(i) the model is arbitrage free if and only if there is a unique equivalent martingale measure (EMM);

(ii) the EMM, if it exists, is the unique probability measure such that the discounted asset price is a martingale;

(iii) any contingent claim has a unique value consistent with the absence of arbitrage;

  (iv) this value is the initial investment required to replicate the claim by trading in the market; and

  (v) the value can be expressed as the expectation of the discounted exercise value, calculated under the EMM.

The question now is whether these five statements continue to hold for other, more realistic, market models. Remarkably, they are all true for the Black–Scholes model—and even for some generalizations of the Black–Scholes model—but it took some years for the full picture to emerge.

BLACK–SCHOLES AND BEYOND

Viewed from a latter-day perspective, the great paper of Black and Scholes of 1973 (hereafter B&S) comes as a huge breath of fresh air. They clear away all the clutter and get straight to the point, announcing their manifesto in the first two sentences of the abstract:

> If options are correctly priced in the market, it should not be possible to make sure profits by creating portfolios of long and short positions in options and their underlying stocks. Using this principle, a theoretical valuation formula for options is derived.

They make what are now the standard assumptions of frictionless markets and adopt Samuelson's geometric Brownian motion price model (3.3) (no dividends are paid). They only consider European options. Then the value of a call option with strike K and exercise time T should just depend on the current time t and underlying price S_t, i.e. it should be equal to $w(S_t, t)$ for some function $w(x, t)$. They recognize immediately that the way to form riskless portfolios is 'delta hedging'. Suppose that $S_t = x$ and that we form a portfolio consisting of one unit of the underlying asset and $-(1/w_1(x, t))$ units of the option, where $w_1 = \partial w/\partial x$. The value of the portfolio is $p = x - w/w_1$ and the change in value in a short time Δt is $\Delta p = \Delta x - \Delta w/w_1$. Expanding w by the Itô formula gives

$$\Delta w = w_1 \Delta x + \tfrac{1}{2} w_{11} \sigma^2 x^2 \Delta t + w_2 \Delta t$$

and hence

$$\Delta p = -\frac{1}{w_1}(\tfrac{1}{2} w_{11} \sigma^2 x^2 + w_2)\Delta t. \tag{3.8}$$

Since this return is certain, it must coincide with the return on the riskless asset, i.e.

$$\Delta p = r p \Delta t = r\left(x - \frac{w}{w_1}\right)\Delta t. \tag{3.9}$$

Equating the right-hand sides of (3.8) and (3.9) and cancelling Δt gives the Black–Scholes PDE

$$w_2 + w_1 r x + \tfrac{1}{2} w_{11} \sigma^2 x^2 - r w = 0, \tag{3.10}$$

and of course we know the value at $t = T$, namely $w(T, x) = \max(x - K, 0)$. B&S introduce a change of variables which reduces (3.10) to the standard heat equation with constant coefficients, which they can solve in integral form, giving the famous formula

$$w(x, t) = x N(d_1) - K e^{-r(T-t)} N(d_2), \tag{3.11}$$

where N is the cumulative normal distribution function and

$$d_1 = \frac{\log(x/K) + (r + \tfrac{1}{2}\sigma^2)(T - t)}{\sigma \sqrt{T - t}}, \qquad d_2 = d_1 - \sigma \sqrt{T - t}.$$

Curiously, they obtain the corresponding value $u(x, t)$ for a put option by noting that the difference $w - u$ satisfies (3.10) with the boundary condition $w(x, T) - u(x, T) = x - K$, to which the solution is easily seen to be

$$w(x, t) - u(x, t) = x - e^{-r(T-t)} K. \tag{3.12}$$

In fact this equality, known as 'put–call parity', is much more fundamental than the B&S formula itself. It follows directly from an arbitrage argument[16] and is not connected with the B&S price model, or indeed any price model.

B&S do not directly use the ideas of replication and the law of one price. Rather, they argue that if the option price function fails to satisfy (3.10), then there is arbitrage derived from constructing a second riskless asset having a growth rate not equal to that of the riskless asset in the market. In this sense they are replicating the riskless asset rather than the option, and the option formula comes out as a by-product. By today's standards their argument is rather informal. They do not give a formal definition of what a trading strategy is, or confirm that their hedging portfolio is self-financing. Of course, these things can be fixed—and soon were by other authors—but the informal approach limited the distance B&S were able to go. For example, there is a paragraph in the paper about the effects of dividends, but they gave no quantitative treatment of this subject. The

---

[16]Being long a call option and short a put option is equivalent to holding a forward contract; the values of these two equivalent positions are, respectively, the left- and right-hand sides of (3.12). This put–call parity was evident to Bachelier who explains (see p. 21) that in speculation on assets one obtains a put (or 'downside option') by buying a call and selling a forward.

fact remains, however, that the B&S paper is the decisive breakthrough in the subject. Any history of option pricing—or of financial economics generally—divides in black-and-white terms into the pre-Black–Scholes and post-Black–Scholes eras.

Regrettably, Fischer Black did not live to collect the economics Nobel Prize for this extraordinary achievement. The 1997 prize[17] was awarded to Myron Scholes and Robert Merton 'for a new method to determine the value of derivatives'. Merton's own paper on option pricing (Merton 1973) contains a cornucopia of good ideas, most of which have now found their way into the standard theory. He introduces a more formal concept of trading strategies and self-financing portfolios. Perhaps most significantly, he allows for stochastic interest rates, introducing as a second asset the zero coupon bond $p(t, T)$ (the value at time t of $1 delivered at time $T \geqslant t$), which is assumed to satisfy

$$\mathrm{d}p(t, T) = \mu(t, T)p(t, T)\,\mathrm{d}t + \delta(t, T)p(t, T)\,\mathrm{d}z_T(t),$$

where μ, δ are deterministic functions and $z_T(t)$ is another Brownian motion. Thus p, as well as the asset price S, has a lognormal distribution. The option price $w(x, p, t)$ must now depend on the current prices $x = S(t)$ and $p = p(t, T)$ of both assets. Merton chooses a 'delta-hedging' strategy in both assets to obtain a portfolio that is 'locally riskless' and obtains a PDE in three variables for the option price w. He now observes that w has the homogeneity property that $w(\lambda x, \lambda p, t) = \lambda w(x, p, t)$ for $\lambda > 0$. Taking $\lambda = 1/p$ we obtain

$$w(x, p, t) = pw\left(\frac{x}{p}, 1, t\right),$$

expressing the option value in terms of a two-parameter function $h(y, t) \equiv w(y, 1, t)$, which is shown to satisfy the B&S PDE with a modified volatility term. This of course reduces to standard B&S when the zero-coupon bond has no volatility, i.e. $\delta(t, T) = 0$.

The significance of this development, apart from the introduction of interest-rate volatility, is the realization that prices are ratios, i.e. should be expressed in units of some *numéraire asset* (in this case, the zero-coupon bond p). Interest-rate modelling soon took off as a subject in its own right, and the numéraire idea is now commonplace in the option pricing literature. Later studies by Jamshidian (1997) and others have

---

[17]Its full name is 'The Bank of Sweden Prize in Economic Sciences in Memory of Alfred Nobel'.

systematically exploited homogeneity properties of price functions. The germs of all these ideas are in Merton's paper.

The next big step forward was taken by Cox and Ross (1976). They give a proof of Black–Scholes which is pretty close to what is found nowadays in dozens of textbooks. Interestingly, though, they extend the argument to cover a limited class of jump-diffusion processes, that is, price processes in which the input 'noise' is Brownian motion plus a Poisson-like jump process. The price process is therefore discontinuous, but Cox and Ross show that nevertheless perfect hedging can be achieved in certain circumstances. They are also able to handle assets paying a dividend yield, a topic with which, as we pointed out above, Black and Scholes had had some difficulty. The main significance of the Cox–Ross paper, however, lies in the introduction of 'risk-neutral valuation'. They note that valuation is preference independent. Therefore it makes no difference what investor preferences are assumed. If we assume that investors are risk neutral, then economic equilibrium requires that the expected returns of all assets be equal to the riskless rate. Option prices can be expressed as discounted expectations in this 'risk-neutral world'. This key idea, specialized to the single-period binomial model, gives the pricing formula (3.7) that we derived above.

The binomial model itself was introduced by Cox and Ross again, together with their collaborator Mark Rubinstein, in 1979. Again, from today's perspective, there is a quite noticeable shift in style in this paper away from 'economics' and towards what is now known as 'financial engineering': an accent on computable models and effective practical techniques. And it is certainly true that the binomial tree, introduced in this paper, soon became, and to some extent remains today, one of the widely used workhorses of the trading environment.

The binomial tree is a discrete-time process consisting of n independent steps of the sort analysed above. A three-period tree is shown in Figure 3.2. Starting at any node, the price moves up by a factor $u > 1$ or down by a factor $d < 1$. By taking $d = 1/u$ and normalizing to an initial price $S_0 = 1$ we get a 'recombining tree' with the prices at the nodes as shown in the figure. As in the single-period model described earlier, we assume that the per-period riskless interest rate is r. The condition for absence of arbitrage is then $d < R < u$ as before, where $R = 1 + r$. Define $h = \log u$, so that $u = e^h$, $d = e^{-h}$, and let $Z_1, Z_2, \ldots$ be i.i.d. random variables with

$$\mathbb{P}[Z_k = 1] = \mathbb{P}[Z_k = -1] = q = (R - d)/(u - d).$$

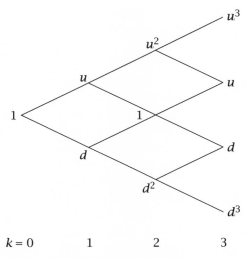

$$k = 0 \qquad 1 \qquad 2 \qquad 3$$

Figure 3.2. Three-period binomial tree.

Then the price process can be expressed as

$$S_k = S_0 \exp\left(h \sum_{i=1}^{k} Z_i\right),$$

and $\mathbb{P}$ is the risk-neutral measure in that the discounted price process $\tilde{S}_k = S_k/R^k$ is a martingale. The process S_k is a 'multiplicative random walk'. If we have an option whose exercise value at time n is $f(S_n)$, then its unique arbitrage-free value at time 0 is the discounted expectation $w(S_0, 0) = R^{-n}\mathbb{E}[f(S_n)]$. This is proved by constructing a hedging portfolio by a direct extension of the one-period case. Further, the expectation may be calculated by a backwards recursion algorithm: if $w(s, k)$ denotes the value of the option at time k when the current price is $S_k = s$, then of course $w(s, n)$ is the exercise value $f(s)$, while, for $k < n$,

$$w(s, k) = \frac{1}{R} E_{(s,k)}[w(S_{k+1}, k+1)]$$

$$= \frac{1}{R}\{qw(us, k+1) + (1-q)w(ds, k+1)\}. \qquad (3.13)$$

At any point (s, k) we face a single-period hedging problem in which the two possible option values at the end of the period are $w(us, k+1)$, $w(ds, k+1)$. From (3.5), the number of units of the risky asset in the replicating portfolio is therefore

$$N(s, k) = \frac{w(us, k+1) - w(ds, k+1)}{(u-d)s},$$

the discrete version of the delta hedge $N = \partial w / \partial s$. Thus the whole pricing and hedging problem is solved simply by computing the backwards recursion (3.13).

Also included in Cox et al. (1979) is the scaling argument that leads to the use of the binomial tree as an approximation to Black–Scholes. The idea is a minor variant on the standard approximation of Brownian motion by a random walk. We take a fixed time interval $[0, T]$ and let the price at step k in an n-period binomial tree represent the asset price S_t, where $t = kT/n$. The time step is thus $\Delta t = T/n$. If we now take the spatial step h as $h = \sigma\sqrt{\Delta t}$ for some constant σ, we find that $X_n(T) := h \sum_1^n Z_i$ has mean μ_n and variance v_n such that $\mu_n \to (r - \frac{1}{2}\sigma^2)T$ and $v_n \to \sigma^2 T$ as $n \to \infty$. Since the Z_i are independent, the Central Limit Theorem implies that the distribution of $X_n(T)$ converges to the normal distribution with the limiting mean and variance. In particular, the value of an option whose exercise value is a continuous function f of the asset price at time T converges to

$$v_0(S_0) = \frac{e^{-rT}}{\sqrt{2\pi}} \int_{-\infty}^{\infty} f(S_0 e^{(r-\sigma^2/2)T+\sigma\sqrt{T}x}) e^{-x^2/2} \, dx,$$

which is equal to the Black–Scholes value. Thus the multiplicative random walk, suitably scaled, approximates geometric Brownian motion, and the binomial tree is in fact providing a simple algorithm for solving the PDE (3.10). This is why the binomial model has had such an enormous impact on practical option trading.

Another reason for the impact of the binomial model is that, as already pointed out in Cox et al. (1979), we can also use it to price American options. The idea is as follows. The right-hand side of (3.13) gives the 'continuation value' of the option as seen at time k. In the European case, this *is* the value since one has no choice but to wait until the final time n before exercising. But in the American case, the holder could exercise *now*, at time k, getting the exercise value $f(s)$. Rationally, he will do this if $f(s)$ is greater than the continuation value, leading to the following relationship between the American values $w_a(\cdot, k)$ and $w_a(\cdot, k+1)$:

$$w_a(s, k) = \max\left\{f(s), \frac{1}{R}\{qw_a(us, k+1) + (1-q)w_a(ds, k+1)\}\right\}. \quad (3.14)$$

On the other hand, at time n the option *must* be exercised, giving $w_a(n, s) = f(s)$. We can therefore generate the American price by backwards recursion using (3.14) together with the same time-n terminal exercise value as in the European case. From (3.13) it is clear that $w_a(s, k) \geqslant f(s)$ and that it is optimal to exercise at time s if $w_a(s, k) = f(s)$. Regarding the binomial model as an approximation to the geometric Brownian motion model, this algorithm, when applied to the put option $f(s) = \max(K - s, 0)$, can be seen as a way of computing the solution to the free-boundary problem introduced by Samuelson (1965).

All of this is contained in the Cox, Ross and Rubinstein paper, and all of it is absolutely correct, but the argument given in the paper is far from complete. It is not at all obvious why the investor should compare the immediate exercise value with the *risk-neutral* expectation of the continuation value. Some separate argument is required to show that this procedure leads to a unique arbitrage-free price. Not until the work of Bensoussan (1984) and Karatzas (1988), some twenty years after McKean's original work, was the connection between free-boundary problems and a unique arbitrage-free value for the American option established in a mathematically watertight way.

Cox and Ross (1976) had arrived at the idea of 'risk-neutral valuation' by observing that the Black–Scholes formula (3.11) does not involve the growth rate α of the stock (3.3), and therefore the value is the same whatever the growth rate. If investors are 'risk neutral', they will assume that the growth rates of all assets, risky or not, is equal to the riskless rate r. But if we take $\alpha = r$ in the Black–Scholes model then the option value given by (3.11) is equal to the discounted expectation $\mathbb{E}[e^{-r(T-t)} \max(S_T - K, 0)]$. Cox and Ross did not, however, make the connection with equivalent martingale measures. This final piece of the jigsaw puzzle was inserted by Harrison and Kreps (1979).

Two probability measures $\mathbb{P}$, $\mathbb{Q}$ on the same probability space $(\Omega, \mathcal{F})$ are *equivalent* if they have the same null sets: for any measurable set A we have $\mathbb{P}[A] = 0$ if and only if $\mathbb{Q}[A] = 0$. When $(\Omega, \mathcal{F})$ is the standard probability space for Brownian motion and $\mathbb{P}$ is Wiener measure, there is a striking characterization of equivalent measures, uncovered by Girsanov (1960). This is that, for Brownian motion $w(t)$, *change of measure is change of drift*, i.e. under any equivalent measure $\mathbb{Q}$ there is a 'drift' process $\phi(t)$ such that $\tilde{w}(t)$, defined by

$$\tilde{w}(t) = w(t) - \int_0^t \phi(s)\, ds,$$

is a Brownian motion under measure $\mathbb{Q}$. More intuitively, we can say that, under $\mathbb{Q}$, $dw(t) = d\tilde{w}(t) + \phi(t)\, dt$, so $w(t)$ is 'Brownian motion plus drift'. If we look at the geometric Brownian motion (3.3), we see that by taking $\phi(t) = (r - \alpha)/\sigma$ we obtain

$$dS(t) = rS(t)\,dt + \sigma S(t)\,d\tilde{w}(t),$$

so that under $\mathbb{Q}$ the price process $S(t)$ is geometric Brownian motion with the risk-neutral growth rate r. We call $\mathbb{Q}$ the equivalent martingale measure (EMM) because under $\mathbb{Q}$ the discounted price process $e^{-rt}S(t)$ is a martingale. The Black–Scholes formula is expressed as $\mathbb{E}_{\mathbb{Q}}[e^{-r(T-t)}\max(S_T - K, 0)]$, where $\mathbb{E}_{\mathbb{Q}}$ denotes expectation under measure $\mathbb{Q}$—exactly the formula (3.7) obtained in the binomial model. Thus, instead of appealing to the economic concept of risk neutrality, we can simply state that the option value is the discounted expectation of the exercise value under the unique EMM.

The above idea is contained in the paper by Harrison and Kreps (1979), but they start from a much more abstract point of view, initially in a single-period setting similar to that of the single-period binomial model discussed above, but allowing for a very general set of random outcomes at the end of the period. Agents in the market have available 'consumption bundles' $(r, x) \in \mathbb{R} \times \mathbb{X}$, where $\mathbb{X}$ is a linear space of random variables. Thus an agent will consume a fixed amount r at time 0 and a random amount $X(\omega)$ at time 1, for some $X \in \mathbb{X}$. A price system is a pair $(\mathbb{M}, \pi)$ where $\mathbb{M}$ is a subspace of $\mathbb{X}$ and π is a linear functional on $\mathbb{M}$. The price of a bundle $(r, m) \in \mathbb{R} \times \mathbb{M}$ is then $r + \pi(m)$. The price system is *viable* if there is an optimal net trade (a bundle such that $r + \pi(m) = 0$) with respect to some preference ordering. The pricing problem is to extend π to all of $\mathbb{X}$ in such a way that the extended market is still viable. A specific claim $X \in \mathbb{X}$ is *priced by arbitrage* if there is a unique price p for X which is consistent with $(\mathbb{M}, \pi)$. The main general results are that

- $(\mathbb{M}, \pi)$ is viable if and only if there is an extension of π to all of $\mathbb{X}$ which is continuous and strictly positive;
- a claim X is priced by arbitrage if and only if it has the same value for all such extensions.

Moving to a multi-period or continuous-time setting where the time interval is $[0, T]$, Harrison and Kreps consider a vector process $S(t)$ of asset prices, and restrict their investors to simple self-financing trading strategies (see Chapter 1 for a description). Starting at time 0 with a certain endowment and applying a trading strategy leads to a random outcome X at time T, to which the single-period theory can be applied. The idea is that $(\mathbb{M}, \pi)$ represents the market of traded assets, while the bigger set $\mathbb{X}$ includes possibly non-marketed contingent claims. The main result is essentially that viability is equivalent to existence of an EMM $\mathbb{Q}$. There is a one-to-one correspondence between EMMs $\mathbb{Q}$ and pricing functionals ψ, given by $\psi(X) = \mathbb{E}_{\mathbb{Q}}X$ and $\mathbb{Q}A = \psi(\mathbf{1}_A)$. The value of X is determined by arbitrage if $\mathbb{E}_{\mathbb{Q}}X$ is the same for every EMM $\mathbb{Q}$. Harrison and Kreps apply this result to finite models similar to the binomial model, and to the Black–Scholes set-up where they bring in the Girsanov theorem to effect the measure change. The paper therefore gives the EMM idea in a variety of settings, but not in any great generality because of the small class of trading strategies to which they restrict themselves.

Harrison continued the quest with co-author Pliska in a further paper in 1981, in which the connection with modern stochastic analysis—the *théorie générale des processus*—was firmly established. This time asset prices are semimartingales and trading strategies are the integrands of the *théorie générale*. They do not directly attack the question of necessary and sufficient conditions for the absence of arbitrage but, rather, assume the existence of an EMM (which is a sufficient condition). They show that the price π of an attainable claim X is always given by $\pi = \mathbb{E}_{\mathbb{Q}}[e^{-rT}X]$. They recognize that a market is 'complete' (all claims are attainable) if the vector process of asset prices has the martingale representation property. By this criterion the Black–Scholes model is complete, and other models will be complete only if they satisfy general conditions given by Jacod and Yor (1977). This paper has turned 'financial economics' into 'mathematical finance'. All questions are posed in purely mathematical terms, and no economic principles that do not have a precise mathematical statement appear.

We leave the story in 1981. It is hard to argue with the view that the period 1965–1981 was the 'heroic period' in financial economics. In 1965, Samuelson was in the middle of his 'desultory researches'; several approaches to option pricing, including Bachelier's, were in the field, but none of them had solid justification; there was no organized market in options to provide reliable data. By 1981 the whole story of complete markets, replicating strategies and equivalent martingale measures, was laid out to view, and this story supported an industry which had already grown to major proportions.

This does not mean, however, that there was nothing left to do. In particular, the exact relation between absence of arbitrage and existence of equivalent martingale measures, so clear in the case of finite probability spaces, was already recognized by Harrison and Pliska to be a delicate question in their general setting. It was more than another decade before this question was satisfactorily resolved, by Delbaen and Schachermayer (1994). Another obvious item on the 'to do' list was American options, the theory of which was settled by Bensoussan (1984) and Karatzas (1988).

In the early days of listed option trading it appeared that the Samuelson–Black–Scholes lognormal diffusion model gave an adequate representation of asset prices. However, there was a sudden 'regime shift' immediately after the October 1987 market crash that has persisted ever since. Post-1987, we observe a 'volatility skew': out-of-the-money put options are quoted at prices corresponding to higher volatilities in the Black–Scholes model than at-the-money or in-the-money puts. This phenomenon led to a concerted effort, which continues to this day, to develop more accurate models to capture the volatility skew and to develop better hedging strategies. A good introduction to these models, usually grouped under the heading 'stochastic volatility', can be found in a collection of papers edited by Lipton (2003).

Besides these topics, the theory has expanded in all kinds of ways. A good idea of the current state of play can be gained by consulting modern textbooks: for example, Karatzas and Shreve (2001), Musiela and Rutkowski (2004) or Föllmer and Schied (2004). Models for interest rates, foreign exchange, commodity options and credit risk have been introduced. The impact of market frictions such as transaction costs has been studied. There have been major investigations of optimal investment and utility maximization, and their relation to pricing and hedging problems in incomplete markets. Walking around exotic derivative trading floors these days, you are just as likely to find Karatzas and Shreve's stochastic calculus bible *Brownian Motion and Stochastic Calculus* lying on the desk as those other two practitioner bibles Hull's *Options, Futures and other Derivatives* and Press et al.'s *Numerical Recipes*. But it is not just the specialists who are involved. Options thinking pervades the entire finance industry. The role of mathematical models in pricing, hedging and risk management is universally appreciated, and any chief executive, risk manager or trader will happily engage in informed discussions of implied volatility, convexity adjustments, gamma risk and a whole host of similar topics, all of which are offshoots of the Black–Scholes theory. In his most extravagant moments Louis Bachelier could hardly have imagined that the seed he planted would grow into a tree this size.

Facsimile of Bachelier's Original Thesis

On the following pages you will find a facsimile of Bachelier's original thesis, first published in 1900 in *Annales Scientifique de l'École Normale Supérieure*, 3e série, tome 17, pp. 21–86.

   This is reprinted here with the permission of École Normale Supérieure de Paris and was scanned for this publication by the Cambridge University Library.

THÉORIE

LA SPÉCULATION,

Par M. L. BACHELIER.

———❦———

INTRODUCTION.

Les influences qui déterminent les mouvements de la Bourse sont innombrables, des événements passés, actuels ou même escomptables, ne présentant souvent aucun rapport apparent avec ses variations, se répercutent sur son cours.

A côté des causes en quelque sorte naturelles des variations, interviennent aussi des causes factices : la Bourse agit sur elle-même et le mouvement actuel est fonction, non seulement des mouvements antérieurs, mais aussi de la position de place.

La détermination de ces mouvements se subordonne à un nombre infini de facteurs : il est dès lors impossible d'en espérer la prévision mathématique. Les opinions contradictoires relatives à ces variations se partagent si bien qu'au même instant les acheteurs croient à la hausse et les vendeurs à la baisse.

Le Calcul des probabilités ne pourra sans doute jamais s'appliquer aux mouvements de la cote et la dynamique de la Bourse ne sera jamais une science exacte.

Mais il est possible d'étudier mathématiquement l'état statique du marché à un instant donné, c'est-à-dire d'établir la loi de probabilité des variations de cours qu'admet à cet instant le marché. Si le marché, en effet, ne prévoit pas les mouvements, il les considère comme étant plus ou moins probables, et cette probabilité peut s'évaluer mathématiquement.

La recherche d'une formule qui l'exprime ne paraît pas jusqu'à ce jour avoir été publiée; elle sera l'objet de ce travail.

J'ai cru nécessaire de rappeler d'abord quelques notions théoriques relatives aux opérations de bourse en y joignant certains aperçus nouveaux indispensables à nos recherches ultérieures.

LES OPÉRATIONS DE BOURSE.

Opérations de bourse. — Il y a deux sortes d'opérations à terme :

Les opérations fermes;

Les opérations à prime.

Ces opérations peuvent se combiner à l'infini, d'autant que l'on traite souvent plusieurs sortes de primes.

Opérations fermes. — Les opérations à terme fermes sont absolument analogues à celles du comptant, mais on règle seulement des différences à une époque fixée d'avance et appelée *liquidation*. Elle a lieu le dernier jour de chaque mois.

Le cours établi le jour de la liquidation et auquel on rapporte toutes les opérations du mois est le *cours de compensation*.

L'acheteur ferme ne limite ni son gain ni sa perte, il gagne la différence entre le prix d'achat et le prix de vente, si la vente est faite au-dessus du prix d'achat, il perd la différence si la vente est faite au-dessous.

Il y a perte pour le vendeur ferme qui rachète plus haut qu'il n'a primitivement vendu, il y a gain dans le cas contraire.

Reports. — L'acheteur au comptant touche ses coupons et peut conserver indéfiniment ses titres. Une opération à terme expirant à la liquidation, l'acheteur à terme doit, pour conserver sa position jusqu'à la liquidation suivante, payer au vendeur une indemnité dite *report* (¹).

(¹) Pour la définition complète des reports, je renvoie aux Ouvrages spéciaux.

Le report varie à chaque liquidation ; sur la rente il est en moyenne de o^{fr},18 par 3^{fr}, mais il peut être plus élevé ou nul ; il peut même être négatif, on l'appelle alors *déport ;* dans ce cas, le vendeur indemnise l'acheteur.

Le jour du détachement du coupon, l'acheteur à terme reçoit du vendeur le montant de ce coupon. En même temps, le cours baisse d'une somme égale ; acheteur et vendeur se trouvent donc immédiatement après le détachement du coupon dans la même position relative qu'avant cette opération.

On voit que si l'acheteur a l'avantage de toucher les coupons, par contre, il doit en général payer des reports. Le vendeur, au contraire, touche les reports, mais il paye les coupons.

Rentes reportables. — Sur la rente, le coupon de o^{fr},75 par trimestre représente o^{fr},25 par mois, alors que le report est presque toujours inférieur à o^{fr},20. La différence est donc à l'avantage de l'acheteur ; de là est venue l'idée d'acheter des rentes pour les faire reporter indéfiniment.

Cette opération est dite de la *rente reportable ;* nous étudierons plus loin ses probabilités de réussite.

Cours équivalents. — Pour bien nous rendre compte du mécanisme des coupons et des reports, faisons abstraction de toutes les autres causes de variation des cours.

Puisque tous les trois mois sur la rente au comptant est détaché un coupon de o^{fr},75 représentant l'intérêt de l'argent de l'acheteur, la rente au comptant doit logiquement monter chaque mois de o^{fr},25. Au cours actuellement coté correspond un cours qui, dans trente jours, serait plus élevé de o^{fr},25, dans quinze jours de o^{fr},125, etc.

Tous ces cours peuvent être considérés comme *équivalents.*

La considération des cours équivalents est beaucoup plus compliquée lorsqu'il s'agit d'opérations à terme. Il est d'abord évident que si le report est nul, le terme doit se comporter comme le comptant et que le cours doit logiquement monter de o^{fr},25 par mois.

Considérons maintenant le cas où le report serait de o^{fr},25. Prenons

119

l'axe des x pour représenter les temps (*fig.* 1), la longueur OA représente un mois compris entre deux liquidations dont l'une correspond au point O et l'autre au point A.

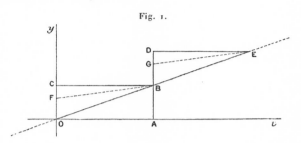

Fig. 1.

Les ordonnées représentent les cours.

Si AB équivaut à $0^{fr}, 25$, la marche logique de la rente au comptant sera représentée par la ligne droite OBE (¹).

Considérons maintenant le cas où le report serait de $0^{fr}, 25$. Un peu avant la liquidation, le comptant et le terme seront au même cours O ; puis, l'acheteur à terme devant payer $0^{fr}, 25$ de report, le cours du terme sautera brusquement de O en C et suivra pendant tout le mois la ligne horizontale CB. En B, il se confondra de nouveau avec le cours du comptant pour augmenter tout à coup de $0^{fr}, 25$ en D, etc.

Dans le cas où le report est une quantité donnée correspondant à la longueur OF, le cours devrait suivre la ligne FB, puis GE, et ainsi de suite. La rente à terme doit donc logiquement dans ce cas, d'une liquidation à l'autre, monter d'une quantité représentée par FC qu'on pourrait appeler le *complément du report*.

Tous les cours de F à B de la ligne FB sont *équivalents* pour les différentes époques auxquelles ils correspondent.

En réalité, l'écart entre le terme et le comptant ne se détend pas d'une façon absolument régulière et FB n'est pas une droite, mais la construction qui vient d'être faite au début du mois peut se répéter à une époque quelconque représentée par le point N.

(¹) On suppose qu'il n'y a pas de détachement de coupon dans l'intervalle considéré, ce qui d'ailleurs ne changerait rien à la démonstration.

Soit NA le temps qui s'écoulera entre l'époque N considérée et la liquidation représentée par le point A.

Fig. 2.

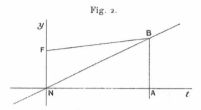

Pendant le temps NA, la rente au comptant doit logiquement monter de AB, proportionnel à NA. Soit NF l'écart entre le comptant et le terme, tous les cours correspondant à la ligne FB sont *équivalents*.

Cours vrais. — Nous appellerons *cours vrai* correspondant à une époque le cours équivalent correspondant à cette époque.

La connaissance du cours vrai a une très grande importance, je vais étudier comment on le détermine.

Désignons par b la quantité dont doit logiquement monter la rente dans l'intervalle d'une journée. Le coefficient b varie généralement peu, sa valeur chaque jour peut être exactement déterminée.

Supposons que n jours nous séparent de la liquidation, et soit C l'écart du terme au comptant.

En n jours, le comptant doit monter de $\frac{25\,n}{30}$ centimes, le terme étant plus élevé de la quantité C ne doit monter pendant ces n jours que de la quantité $\frac{25\,n}{30} - C$, c'est-à-dire pendant un jour de

$$\frac{1}{n}\left(\frac{25\,n}{30} - C\right) = \frac{1}{6\,n}(5\,n - 6\,C).$$

On a donc

$$b = \frac{1}{6\,n}(5\,n - 6\,C).$$

La moyenne des cinq dernières années donne

$$b = 0^c,264.$$

Le cours vrai correspondant à *m* jours sera égal au cours coté actuellement, augmenté de la quantité *mb*.

Représentation géométrique des opérations fermes. — Une opération peut se représenter géométriquement d'une façon très simple, l'axe des *x* représentant les différents cours et l'axe des *y* les bénéfices correspondants.

Je suppose que j'aie fait un achat ferme au cours représenté par O, que je prends pour origine. Au cours $x = OA$, l'opération donne pour

Fig. 3.

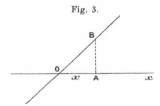

bénéfice *x*; et comme, l'ordonnée correspondante doit être égale au bénéfice, $AB = OA$; l'achat ferme est donc représenté par la ligne OB inclinée à 45° sur la ligne des cours.

Une vente ferme se représenterait d'une façon inverse.

Primes. — Dans l'achat ou la vente ferme, acheteurs et vendeurs s'exposent à une perte théoriquement illimitée. Dans le marché à prime, l'acheteur paye le titre plus cher que dans le cas du marché ferme, mais sa perte en baisse est limitée d'avance à une certaine somme qui est le montant de la prime.

Le vendeur de prime a l'avantage de vendre plus cher, mais il ne peut avoir pour bénéfice que le montant de la prime.

On fait aussi des primes à la baisse qui limitent la perte du vendeur; dans ce cas, l'opération se fait à un cours inférieur à celui du ferme.

On ne traite ces primes à la baisse que dans la spéculation sur les marchandises; dans la spéculation sur les valeurs, on obtient une prime à la baisse en vendant ferme et en achetant simultanément à prime. Pour fixer les idées, je ne m'occuperai que des primes à la hausse.

Supposons, par exemple, que le 3 % cote 104^{fr} au début du mois; si nous en achetons 3000 ferme, nous nous exposons à une perte qui peut devenir considérable s'il se produit une forte baisse.

Pour éviter ce risque, nous pouvons acheter une prime dont 50^c ([1]) en payant, non plus 104^{fr}, mais 104^{fr},15, par exemple; notre cours d'achat est plus élevé, il est vrai, mais notre perte reste limitée quelle que soit la baisse à 50^c par 3^{fr}, c'est-à-dire à 500^{fr}.

L'opération est la même que si nous avions acheté du ferme à 104^{fr},15, ce ferme ne pouvant baisser de plus de 50^c, c'est-à-dire descendre au-dessous de 103^{fr},65.

Le cours de 103^{fr},65, dans le cas actuel, est le *pied de la prime*.

On voit que le cours du pied de la prime est égal au cours auquel elle est négociée, diminué du montant de la prime.

Réponse des primes. — La veille de la liquidation, c'est-à-dire l'avant-dernier jour du mois, a lieu la *réponse des primes*. Reprenons l'exemple précédent et supposons qu'à cet instant de la réponse le cours de la rente soit inférieur à 103^{fr},65, nous *abandonnerons* notre prime, qui sera le bénéfice de notre vendeur.

Si, au contraire, le cours de la réponse est supérieur à 103^{fr},65, notre opération sera transformée en opération ferme; on dit dans ce cas que la prime est *levée*.

En résumé, une prime est levée ou abandonnée suivant que le cours de la réponse est inférieur ou supérieur au pied de la prime.

On voit que les opérations à prime ne courent pas jusqu'à la liquidation; si la prime est levée à la réponse, elle devient du ferme et se liquide le lendemain.

Dans tout ce qui suivra, nous supposerons que le cours de compensation se confond avec le cours de la réponse des primes; cette hypothèse est justifiable, car rien n'empêche de liquider ses opérations à la réponse des primes.

([1]) On dit *une prime dont* pour *une prime de* et l'on emploie la notation 104,15/50 pour désigner une opération faite au cours de 104^{fr},15 dont 50^c.

Écart des primes. — L'écart entre le cours du ferme et celui d'une prime dépend d'un grand nombre de facteurs et varie sans cesse.

Au même instant, l'écart est d'autant plus grand que la prime est plus faible; par exemple, la prime dont 50^c est évidemment meilleur marché que la prime dont 25^c.

L'écart d'une prime décroît plus ou moins régulièrement depuis le commencement du mois jusqu'à la veille de la réponse, moment où cet écart devient très faible.

Mais, suivant les circonstances, il peut se détendre très irrégulièrement et se trouver plus grand quelques jours avant la réponse qu'au commencement du mois.

Primes pour fin prochain. — On traite des primes non seulement pour fin courant, mais aussi pour fin prochain. L'écart de celles-ci est nécessairement plus grand que celui des primes fin courant, mais il est plus faible qu'on ne le croirait en faisant la différence entre le cours de la prime et celui du ferme; il faut en effet déduire de cet écart apparent l'importance du report fin courant.

Par exemple, l'écart moyen de la prime $/25^c$ à 45 jours de la réponse est en moyenne de 72^c; mais, comme le report moyen est de 17^c, l'écart n'est en réalité que de 55^c.

Le détachement d'un coupon fait baisser le cours de la prime d'une valeur égale à l'importance du coupon. Si, par exemple, j'achète, le 2 septembre, une prime $/25^c$ à $104^{fr}, 5o$ fin courant, le cours de ma prime sera devenu $103^{fr}, 75$ le 16 septembre après le détachement du coupon.

Le cours du pied de la prime sera $103^{fr}, 5o$.

Primes pour le lendemain. — On traite, surtout en coulisse, des primes dont 5^c et quelquefois dont 10^c pour le lendemain.

La réponse pour ces petites primes a lieu tous les jours à 2^h.

Les primes en général. — Dans un marché à prime pour une échéance donnée, il y a deux facteurs à considérer : l'importance de la prime et son écart du ferme.

Il est bien évident que plus une prime est forte, plus son écart est petit.

Pour simplifier la négociation des primes, on les a ramenées à trois types en faisant sur l'importance de la prime et sur son écart les trois hypothèses les plus simples :

1° L'importance de la prime est constante et son écart est variable ; c'est cette sorte de prime qui se négocie sur les valeurs ; par exemple, sur le 3 %, on traite des primes /5o^c, /25^c et /1o^c.

2° L'écart de la prime est constant et son importance est variable ; c'est ce qui a lieu pour les primes à la baisse sur les valeurs (c'est-à-dire la vente ferme contre achat à prime).

3° L'écart de la prime est variable ainsi que son importance, mais ces deux quantités sont toujours égales. C'est ainsi que l'on traite les primes sur les marchandises. Il est évident qu'en employant ce dernier système on ne peut traiter à un moment donné qu'une seule prime pour la même échéance.

Remarque sur les primes. — Nous examinerons quelle est la loi qui régit les écarts des primes ; cependant nous pouvons, dès maintenant, faire une remarque assez curieuse :

Une prime doit être d'autant plus forte que son écart est plus faible. Ce fait évident ne suffit pas pour montrer que l'usage des primes soit rationnel.

J'ai en effet reconnu, il y a plusieurs années, qu'il était possible en l'admettant d'imaginer des opérations où l'un des contractants gagnerait à tous les cours.

Sans reproduire les calculs, élémentaires mais assez pénibles, je me contente de présenter un exemple.

L'opération suivante :

Achat d'une unité /1^{fr},
Vente de quatre unités /5o^c,
Achat de trois unités /25^c,

donnerait un bénéfice à tous les cours pourvu que l'écart du /25^c au /5o^c soit au plus le tiers de l'écart du /5o^c au /1^{fr}.

Nous verrons que des écarts semblables ne se rencontrent jamais dans la pratique.

Représentation géométrique des opérations à prime. — Proposons-nous de représenter géométriquement un achat à primes.

Prenons, par exemple, pour origine le cours du ferme au moment où la prime dont h a été traitée; soit E_1 le cours relatif de cette prime ou son écart.

Au-dessus du pied de la prime, c'est-à-dire au cours $(E_1 - h)$ représenté par le point A, l'opération est assimilable à une opération ferme traitée au cours E_1; elle est donc représentée par la ligne CBF. Au-dessous du cours $E_1 - h$, la perte est constante et, par suite, l'opération est représentée par la ligne brisée DCF.

Fig. 4.

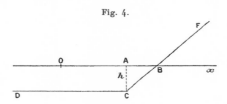

La vente à prime se représenterait d'une façon inverse.

Écarts vrais. — Jusqu'à présent nous n'avons parlé que des écarts cotés, les seuls dont on s'occupe ordinairement; ce ne sont cependant pas eux qui s'introduiront dans notre théorie, mais bien les *écarts vrais*, c'est-à-dire les écarts entre les cours des primes et les cours vrais correspondant à la réponse des primes. Le cours dont il s'agit étant supérieur au cours coté (à moins que le report soit supérieur à 25ᶜ, ce qui est rare), il en résulte que l'écart vrai d'une prime est inférieur à son écart coté.

L'écart vrai d'une prime traitée n jours avant la réponse sera égal à son écart diminué de la quantité nb.

L'écart vrai d'une prime pour fin prochain sera égal à son écart coté diminué de la quantité $[25 + (n - 3o)b]$.

Options. — On traite sur certains marchés des opérations en quelque sorte intermédiaires entre les opérations fermes et les opérations à prime, ce sont les options.

Supposons que 30^{fr} soient le cours d'une marchandise. Au lieu d'acheter une unité au cours de 30^{fr} pour une échéance donnée, nous pouvons acheter une option du double pour la même échéance à 32^{fr}, par exemple. Il faut entendre par là que pour toute différence au-dessous du cours de 32^{fr}, nous ne perdons que sur une unité, alors que pour toute différence au-dessus, nous gagnons sur deux unités.

Nous aurions pu acheter une option du triple à 33^{fr}, par exemple, c'est-à-dire que, pour toute différence au-dessous du cours de 33^{fr} nous perdons sur une unité, alors que pour toute différence au-dessus de ce cours nous gagnons sur trois unités. On peut imaginer des options d'un ordre multiple, la représentation géométrique de ces opérations ne présente aucune difficulté.

On traite aussi des options à la baisse, nécessairement au même écart que les options à la hausse du même ordre de multiplicité.

LES PROBABILITÉS DANS LES OPÉRATIONS DE BOURSE.

Probabilités dans les opérations de bourse. — On peut considérer deux sortes de probabilités :

1° La probabilité que l'on pourrait appeler *mathématique,* c'est celle que l'on peut déterminer *a priori;* celle que l'on étudie dans les jeux de hasard.

2° La probabilité dépendant de faits à venir et, par conséquent, impossible à prévoir d'une façon mathématique.

C'est cette dernière probabilité que cherche a prévoir le spéculateur, il analyse les raisons qui peuvent influer sur la hausse ou sur la baisse et sur l'amplitude des mouvements. Ses inductions sont absolument personnelles, puisque sa contre-partie a nécessairement l'opinion inverse.

Il semble que le marché, c'est-à-dire l'ensemble des spéculateurs, ne doit croire *à un instant donné* ni à la hausse, ni à la baisse, puisque, pour chaque cours coté, il y a autant d'acheteurs que de vendeurs.

En réalité, le marché croit à la hausse provenant de la différence entre les coupons et les reports; les vendeurs font un léger sacrifice qu'ils considèrent comme compensé.

On peut ne pas tenir compte de cette différence, à la condition de considérer les cours vrais correspondant à la liquidation, mais les opérations se réglant sur les cours cotés, le vendeur paye la différence.

Par la considération des cours vrais on peut dire :

Le marché ne croit, à un instant donne, ni à la hausse, ni à la baisse du cours vrai.

Mais, si le marché ne croit ni à la hausse, ni à la baisse du cours vrai, il peut supposer plus ou moins probables des mouvements d'une certaine amplitude.

La détermination de la loi de probabilité qu'admet le marché à un instant donné sera l'objet de cette étude.

L'espérance mathématique. — On appelle *espérance mathématique* d'un bénéfice éventuel le produit de ce bénéfice par la probabilité correspondante.

L'*espérance mathématique totale* d'un joueur sera la somme des produits des bénéfices éventuels par les probabilités correspondantes.

Il est évident qu'un joueur ne sera ni avantagé, ni lésé si son espérance mathématique totale est nulle.

On dit alors que le jeu est *équitable*.

On sait que les jeux de courses et tous ceux qui sont pratiqués dans les maisons de jeu ne sont pas équitables : la maison de jeu ou le donneur s'il s'agit de paris aux courses, jouent avec une espérance positive, et les pontes avec une espérance négative.

Dans ces sortes de jeux les pontes n'ont pas le choix entre l'opération qu'ils font et sa contre-partie; comme il n'en est pas de même à la Bourse, il peut sembler curieux que ces jeux ne soient pas équitables, le vendeur acceptant *a priori* un désavantage si les reports sont inférieurs aux coupons.

L'existence d'une seconde sorte de probabilités explique ce fait qui peut sembler paradoxal.

L'avantage mathématique. — L'espérance mathématique nous indique si un jeu est avantageux ou non : elle nous apprend de plus ce que le jeu doit logiquement faire gagner ou faire perdre ; mais elle ne donne pas un coefficient représentant, en quelque sorte, la valeur intrinsèque du jeu.

Ceci va nous amener à introduire une nouvelle notion : celle de l'avantage mathématique.

Nous appellerons *avantage mathématique* d'un joueur le rapport de son espérance positive à la somme arithmétique de ses espérances positive et négative.

L'avantage mathématique varie comme la probabilité de zéro à un, il est égal à $\frac{1}{2}$ quand le jeu est équitable.

Principe de l'espérance mathématique. — On peut assimiler l'acheteur au comptant à un joueur ; en effet, si le titre peut monter après l'achat, la baisse est également possible. Les causes de cette hausse ou de cette baisse rentrent dans la seconde catégorie de probabilités.

D'après la première le titre (¹) doit monter d'une valeur égale à l'importance de ses coupons ; il en résulte qu'au point de vue de cette première classe de probabilités :

L'espérance mathématique de l'acheteur au comptant est positive.

Il est évident qu'il en sera de même de l'espérance mathématique de l'acheteur à terme si le report est nul, car son opération sera assimilable à celle de l'acheteur au comptant.

Si le report sur la rente était de 25ᶜ, l'acheteur ne serait pas plus avantagé que le vendeur.

On peut donc dire :

Les espérances mathématiques de l'acheteur et du vendeur sont nulles quand le report est de 25ᶜ.

Quand le report est inférieur à 25ᶜ, ce qui est le cas ordinaire :

(¹) Je considère le cas le plus simple d'un titre à revenu fixe, sinon l'augmentation du revenu serait une probabilité de la seconde classe.

L'espérance mathématique de l'acheteur est positive, celle du vendeur est négative.

Il faut toujours remarquer qu'il s'agit uniquement de la première sorte de probabilités.

D'après ce qui a été vu précédemment, on peut toujours considérer le report comme étant de 25^c à la condition de remplacer le cours coté par le cours *vrai* correspondant à la liquidation ; si donc, on considère ces cours vrais on peut dire que :

Les espérances mathématiques de l'acheteur et du vendeur sont nulles.

Au point de vue des reports on peut considérer la réponse des primes comme se confondant avec la liquidation ; donc :

Les espérances mathématiques de l'acheteur et du vendeur de primes sont nulles.

En résumé, la considération des cours vrais permet d'énoncer ce principe fondamental :

L'espérance mathématique du spéculateur est nulle.

Il faut bien se rendre compte de la généralité de ce principe : il signifie que le marché, à un instant donné, considère comme ayant une espérance nulle non seulement les opérations traitables actuellement, mais encore celles qui seraient basées sur un mouvement ultérieur des cours.

Par exemple, j'achète de la rente avec l'intention de la revendre lorsqu'elle aura monté de $5o^c$, l'espérance de cette opération complexe est nulle absolument comme si j'avais l'intention de revendre ma rente en liquidation ou à un moment quelconque.

L'espérance d'une opération ne peut être positive ou négative que s'il se produit un mouvement des cours, *a priori* elle est nulle.

Forme générale de la courbe de probabilité. — La probabilité pour que le cours y soit coté à une époque donnée est une fonction de y.

On pourra représenter cette probabilité par l'ordonnée d'une courbe dont les abscisses correspondront aux différents cours.

Il est évident que le cours considéré par le marché comme le plus probable est le cours vrai actuel : si le marché en jugeait autrement, il coterait non pas ce cours, mais un autre plus ou moins élevé.

Dans la suite de cette étude, nous prendrons pour origine des coordonnées le cours vrai correspondant à l'époque donnée. Le cours pourra varier entre $-x_0$ et $+\infty$; x_0 étant le cours absolu actuel.

Nous supposerons qu'il puisse varier entre $-\infty$ et $+\infty$; la probabilité d'un écart plus grand que x_0 étant considérée *a priori* comme tout à fait négligeable.

Dans ces conditions, on peut admettre que la probabilité d'un écart à partir du cours vrai est indépendante de la valeur absolue de ce cours, et que la courbe des probabilités est symétrique par rapport au cours vrai.

Dans ce qui suivra, il ne sera question que du cours relatif, l'origine des coordonnées correspondra toujours au cours vrai actuel.

La loi de probabilité. — La loi de probabilité peut se déterminer par le principe de la probabilité composée.

Désignons par $p_{x,t}\,dx$, la probabilité pour que, à l'époque t, le cours se trouve compris dans l'intervalle élémentaire x, $x+dx$.

Cherchons la probabilité pour que le cours z soit coté à l'époque t_1+t_2, le cours x ayant été coté à l'époque t_1.

En vertu du principe de la probabilité composée, la probabilité cherchée sera égale au produit de la probabilité pour que le cours x soit coté à l'époque t_1, c'est-à-dire $p_{x,t_1}\,dx$, multipliée par la probabilité pour que, le cours x étant coté à l'époque t_1, le cours z soit coté à l'époque t_1+t_2, c'est-à-dire, multipliée par $p_{z-x,t_2}\,dz$.

La probabilité cherchée est donc

$$p_{x,t_1}\,p_{z-x,t_2}\,dx\,dz.$$

Le cours pouvant se trouver à l'époque t_1 dans tous les intervalles dx compris entre $-\infty$ et $+\infty$, la probabilité pour que le cours z soit coté à l'époque t_1+t_2 sera

$$\int_{-\infty}^{+\infty} p_{x,t_1}\,p_{z-x,t_2}\,dx\,dz.$$

La probabilité de ce cours z, à l'époque t_1+t_2, a aussi pour expression p_{z,t_1+t_2}; on a donc

$$p_{z,t_1+t_2}\,dz = \int_{-\infty}^{+\infty} p_{x,t_1}\,p_{z-x,t_2}\,dx\,dz$$

ou

$$p_{z,t_1+t_2} = \int_{-\infty}^{+\infty} p_{x,t_1} p_{z-x,t_2}\, dx,$$

telle est l'équation de condition à laquelle doit satisfaire la fonction p.

Cette équation est vérifiée, comme nous allons le voir, par la fonction

$$p = A\, e^{-B^2 x^2}.$$

Remarquons dès maintenant que l'on doit avoir

$$\int_{-\infty}^{+\infty} p\, dx = A \int_{-\infty}^{+\infty} e^{-B^2 x^2}\, dx = 1.$$

L'intégrale classique qui figure dans le premier terme a pour valeur $\dfrac{\sqrt{\pi}}{B}$, on a donc $B = A\sqrt{\pi}$ et, par suite

$$p = A\, e^{-\pi A^2 x^2}.$$

En posant $x = 0$, on obtient $A = p_0$, c'est-à-dire : A égale la probabilité du cours coté actuellement.

Il faut donc établir que la fonction

$$p = p_0 e^{-\pi p_0^2 x^2},$$

où p_0 dépend du temps, satisfait bien à l'équation de condition ci-dessus.

Soient p_1 et p_2 les quantités correspondant à p_0 et relatives aux temps t_1 et t_2, il faut prouver que l'expression

$$\int_{-\infty}^{+\infty} p_1 e^{-\pi p_1^2 x^2} \times p_2 e^{-\pi p_2^2 (z-x)^2}\, dx$$

peut se mettre sous la forme $A e^{-B^2 z^2}$; A et B ne dépendant que du temps.

Cette intégrale devient, en remarquant que z est une constante,

$$p_1 p_2 e^{-\pi p_2^2 z^2} \int_{-\infty}^{+\infty} e^{-\pi(p_1^2 + p_2^2) x^2 + 2\pi p_2^2 zx}\, dx$$

ou

$$p_1 p_2 e^{-\pi p_2^2 z^2 + \frac{\pi p_2^4 z^2}{p_1^2 + p_2^2}} \int_{-\infty}^{+\infty} e^{-\pi\left(x\sqrt{p_1^2+p_2^2} - \frac{p_2^2 z}{\sqrt{p_1^2+p_2^2}}\right)^2}\, dx;$$

posons

$$x\sqrt{p_1^2+p_2^2}-\frac{p_2^2 z}{\sqrt{p_1^2+p_2^2}}=u;$$

nous aurons alors

$$\frac{p_1 p_2 e^{-\pi p_2^2 z^2\frac{\pi p_2^4 z^2}{p_1^2+p_2^2}}}{\sqrt{p_1^2+p_2^2}}\int_{-\infty}^{+\infty}e^{-\pi u^2}\,du.$$

L'intégrale ayant pour valeur 1, nous obtenons finalement

$$\frac{p_1 p_2}{\sqrt{p_1^2+p_2^2}}\,e^{-\pi\frac{p_1^2 p_2^2}{p_1^2+p_2^2}z^2}.$$

Cette expression ayant la forme désirée, on doit en conclure que la probabilité s'exprime bien par la formule

$$p=p_0 e^{-\pi p_0^2 x^2},$$

dans laquelle p_0 dépend du temps.

On voit que la probabilité est régie par la loi de Gauss déjà célèbre dans le Calcul des probabilités.

Probabilité en fonction du temps. — La formule antéprécédente nous montre que les paramètres $p_0=f(t)$ satisfont à la relation fonctionnelle

$$f^2(t_1+t_2)=\frac{f^2(t_1)f^2(t_2)}{f^2(t_1)+f^2(t_2)};$$

différentions par rapport à t_1, puis par rapport à t_2. Le premier membre ayant la même forme dans les deux cas, nous obtenons

$$\frac{f'(t_1)}{f^3(t_1)}=\frac{f'(t_2)}{f^3(t_2)}.$$

Cette relation ayant lieu, quels que soient t_1 et t_2, la valeur commune des deux rapports est constante, et l'on a

$$f'(t)=Cf^3(t),$$

d'où

$$f(t)=p_0=\frac{H}{\sqrt{t}},$$

H désignant une constante.

Nous avons donc pour expression de la probabilité

$$p = \frac{H}{\sqrt{t}} e^{-\frac{\pi H^2 x^2}{t}}.$$

Espérance mathématique. — L'espérance correspondant au cours x a pour valeur

$$\frac{H x}{\sqrt{t}} e^{-\frac{\pi H^2 x^2}{t}}.$$

L'espérance positive totale est donc

$$\int_0^\infty \frac{H x}{\sqrt{t}} e^{-\frac{\pi H^2 x^2}{t}} dx = \frac{\sqrt{t}}{2 \pi H}.$$

Nous prendrons pour constante, dans notre étude, l'espérance mathématique k correspondant à $t = 1$; nous aurons donc

$$k = \frac{1}{2 \pi H} \qquad \text{ou} \qquad H = \frac{1}{2 \pi k}.$$

L'expression définitive de la probabilité est donc

$$p = \frac{1}{2 \pi k \sqrt{t}} e^{-\frac{x^2}{4 \pi k^2 t}}.$$

L'espérance mathématique

$$\int_0^\infty p x \, dx = k \sqrt{t}$$

est proportionnelle à la racine carrée du temps.

Nouvelle détermination de la loi de probabilité. — L'expression de la fonction p peut s'obtenir en suivant une voie différente de celle que nous avons employée.

Je suppose que deux événements contraires A et B aient pour probabilités respectives p et $q = 1 - p$. La probabilité pour que, sur m événements, il s'en produise α égaux à A et $m - \alpha$ égaux à B a pour expression

$$\frac{m!}{\alpha!(m-\alpha)!} p^\alpha q^{m-\alpha}.$$

C'est un des termes du développement de $(p + q)^m$.
La plus grande de ces probabilités a lieu pour

$$\alpha = mp \quad \text{et} \quad (m - \alpha) = mq.$$

Considérons le terme dont l'exposant de p est $mp + h$, la probabilité correspondante est

$$\frac{m!}{(mp + h)!\,(mq - h)!}\, p^{mp+h} q^{mq-h}.$$

La quantité h est appelée l'*écart*.

Cherchons quelle serait l'espérance mathématique d'un joueur qui toucherait une somme égale à l'écart quand cet écart serait positif.

Nous venons de voir que la probabilité d'un écart h est le terme du développement de $(p + q)^m$ dans lequel l'exposant de p est $mp + h$ et celui de q, $mq - h$. Pour obtenir l'espérance mathématique correspondant à ce terme, il faudra multiplier cette probabilité par h; or

$$h = q(mp + h) - p(mq - h),$$

$mp + h$ et $mq - h$ sont les exposants de p et de q dans le terme de $(p + q)^m$. Multiplier un terme

$$q^\mu p^\nu$$

par

$$\nu q - \mu p = pq\left(\frac{\nu}{p} - \frac{\mu}{q}\right),$$

c'est prendre la dérivée par rapport à p, en retrancher la dérivée par rapport à q, et multiplier la différence par pq.

Pour obtenir l'espérance mathématique totale, nous devons donc prendre les termes du développement de $(p + q)^m$ pour lesquels h est positif, c'est-à-dire

$$p^m + mp^{m-1}q + \frac{m(m-1)}{1.2}\,p^{m-2}q^2 + \ldots + \frac{m!}{mp!\,mq!}\,p^{mp}q^{mq},$$

et retrancher la dérivée par rapport à q de la dérivée par rapport à p, pour multiplier ensuite le résultat par pq.

La dérivée du second terme par rapport à q est égale à la dérivée du premier par rapport à p, la dérivée du troisième par rapport à q est la dérivée du second par rapport à p, et ainsi de suite. Les termes se détruisent donc deux à deux et il ne reste que la dérivée du dernier par rapport à p

$$\frac{m\,!}{mp\,!\,mq\,!}\,p^{mp}\,q^{mq}\,mpq.$$

La valeur moyenne de l'écart h serait égale au double de cette quantité.

Lorsque le nombre m est suffisamment grand, on peut simplifier les expressions précédentes en faisant usage de la formule asymptotique de Stirling

$$n\,! = e^{-n}\,n^{n}\sqrt{2\pi n}.$$

On obtient ainsi pour l'espérance mathématique la valeur

$$\frac{\sqrt{mpq}}{\sqrt{2\pi}}.$$

La probabilité pour que l'écart h soit compris entre h et $h + dh$ aura pour expression

$$\frac{dh}{\sqrt{2\pi mpq}}\,e^{-\frac{h^2}{2mpq}}.$$

Nous pouvons appliquer la théorie qui précède à notre étude. On peut supposer le temps divisé en intervalles très petits Δt; de sorte que $t = m\,\Delta t$; pendant le temps Δt le cours variera probablement très peu.

Formons la somme des produits des écarts qui peuvent exister à l'époque Δt par les probabilités correspondantes; c'est-à-dire $\int_{0}^{\infty} px\,dx$, p étant la probabilité de l'écart x.

Cette intégrale doit être finie, car, par suite de la petitesse supposée de Δt, les écarts considérables ont une probabilité évanouissante. Cette intégrale exprime du reste une espérance mathématique, qui ne peut être finie si elle correspond à un intervalle de temps très petit.

Désignons par Δx le double de la valeur de l'intégrale ci-dessus; Δx sera la moyenne des écarts ou l'écart moyen pendant le temps Δt.

Si le nombre m des épreuves est très grand et si la probabilité reste la même à chaque épreuve, nous pourrons supposer que le cours varie pendant chacune des épreuves Δt de l'écart moyen Δx; la hausse Δx aura pour probabilité $\frac{1}{2}$, comme aussi la baisse $-\Delta x$.

La formule qui précède donnera donc, en y faisant $p = q = \frac{1}{2}$, la probabilité pour que, à l'époque t, le cours soit compris entre x et $x + dx$, ce sera

$$\frac{2\,dx\sqrt{\Delta t}}{\sqrt{2\pi}\sqrt{t}}\,e^{-\frac{2x^2\Delta t}{t}},$$

ou, en posant $H = \frac{2}{\sqrt{2\pi}}\sqrt{\Delta t}$,

$$\frac{H\,dx}{\sqrt{t}}\,e^{-\frac{\pi H^2 x^2}{t}}.$$

L'espérance mathématique aura pour expression

$$\frac{\sqrt{t}}{2\sqrt{2\pi}\sqrt{\Delta t}} = \frac{\sqrt{t}}{2\pi H}.$$

Si nous prenons pour constante l'espérance mathématique k correspondant à $t = 1$, nous trouvons, comme précédemment,

$$p = \frac{1}{2\pi k\sqrt{t}}\,e^{-\frac{x^2}{4\pi k^2 t}}.$$

Les formules précédentes donnent $\Delta t = \frac{1}{8\pi k^2}$; la variation moyenne pendant cet intervalle de temps est

$$\Delta x = \frac{\sqrt{2}}{2\sqrt{\pi}}.$$

Si nous posons $x = n\,\Delta x$, la probabilité aura pour expression

$$p = \frac{\sqrt{2}}{\sqrt{\pi}\sqrt{m}}\,e^{-\frac{n^2}{\pi m}}.$$

Courbe des probabilités. — La fonction

$$p = p_0\,e^{-\pi p_1^2 x^2}$$

peut se représenter par une courbe dont l'ordonnée est maxima à l'origine et qui présente deux points d'inflexion pour

$$x = \pm \frac{1}{p_0\sqrt{2\pi}} = \pm \sqrt{2\pi}\, k \sqrt{t}.$$

Ces mêmes valeurs de x sont aussi les abscisses des maxima et minima des courbes d'espérance mathématique, dont l'équation est

$$y = \pm p x.$$

La probabilité du cours x est une fonction de t; elle croît jusqu'à une certaine époque et décroît ensuite. La dérivée $\frac{dp}{dt} = 0$ lorsque $t = \frac{x^2}{2\pi k^2}$. La probabilité du cours x est donc maxima quand ce cours correspond au point d'inflexion de la courbe des probabilités.

Probabilité dans un intervalle donné. — L'intégrale

$$\frac{1}{2\pi k \sqrt{t}} \int_0^x e^{-\frac{x^2}{4\pi k^2 t}}\, dx = \frac{c}{\sqrt{\pi}} \int_0^x e^{-c^2 x^2}\, dx$$

n'est pas exprimable en termes finis, mais on peut donner son développement en série

$$\frac{1}{\sqrt{\pi}}\left[cx - \frac{\frac{1}{3}(cx)^3}{1} + \frac{\frac{1}{5}(cx)^5}{1.2} - \frac{\frac{1}{7}(cx)^7}{1.2.3} + \cdots \right].$$

Cette série converge assez lentement pour les valeurs très fortes de cx. Laplace a donné pour ce cas l'intégrale définie sous la forme d'une fraction continue fort aisée à calculer

$$\frac{1}{2} - \frac{e^{-c^2 x^2}}{2cx\sqrt{\pi}} \cfrac{1}{1 + \cfrac{\alpha}{1 + \cfrac{2\alpha}{1 + \cfrac{3\alpha}{1 + \cdots}}}},$$

dans laquelle $\alpha = \frac{1}{2c^2 x^2}$.

138

Les réduites successives sont

$$\frac{1}{1+\alpha}, \quad \frac{1+2\alpha}{1+3\alpha}, \quad \frac{1+5\alpha}{1+6\alpha+3\alpha^2}, \quad \frac{1+9\alpha+8\alpha^2}{1+10\alpha+15\alpha^2}.$$

Il existe un autre procédé permettant de calculer l'intégrale ci-dessus quand x est un grand nombre.

On a

$$\int_x^\infty e^{-x^2}\,dx = \int_x^\infty \frac{1}{2x}\,e^{-x^2}\,2x\,dx;$$

en intégrant par parties, on obtient alors

$$\begin{aligned}
\int_x^\infty e^{-x^2}\,dx &= \frac{e^{-x^2}}{2x} - \int_x^\infty e^{-x^2}\frac{dx}{2x^2} \\
&= \frac{e^{-x^2}}{2x} - \frac{e^{-x^2}}{4x^3} + \int_x^\infty e^{-x^2}\frac{1\cdot3}{4x^4}\,dx \\
&= \frac{e^{-x^2}}{2x} - \frac{e^{-x^2}}{4x^3} + \frac{e^{-x^2}1\cdot3}{8x^5} - \int_x^\infty e^{-x^2}\frac{1\cdot3\cdot5}{8x^6}\,dx.
\end{aligned}$$

Le terme général de la série a pour expression

$$\frac{1\cdot3\cdot5\ldots(2n-1)}{2^{2n-1}x^{2n+1}}\,e^{-x^2}.$$

Le rapport d'un terme au précédent dépasse l'unité lorsque $2n+1 > 4x^2$. La série diverge donc à partir d'un certain terme. On peut obtenir une limite supérieure de l'intégrale qui sert de reste.

On a, en effet,

$$\begin{aligned}
\frac{1\cdot3\cdot5\ldots(2n+1)}{2^{2n-1}}\int_x^\infty \frac{e^{-x^2}}{x^{2n+2}}\,dx &< \frac{1\cdot3\cdot5\ldots(2n+1)}{2^{2n-1}}\,e^{-x^2}\int_x^\infty \frac{dx}{x^{2n+2}} \\
&= \frac{1\cdot3\cdot5\ldots(2n-1)}{2^{2n-1}x^{2n+1}}\,e^{-x^2}.
\end{aligned}$$

Or cette dernière quantité est le terme qui précédait l'intégrale. Le terme complémentaire est donc toujours plus petit que celui qui le précède.

On a édité des tables donnant les valeurs de l'intégrale

$$\Theta(y) = \frac{2}{\sqrt{\pi}}\int_0^y e^{-y^2}\,dy.$$

139

On aura évidemment

$$\int_0^x p\,dx = \frac{1}{2}\,\Theta\left(\frac{x}{2\,k\sqrt{\pi}\sqrt{t}}\right).$$

La probabilité

$$\mathscr{P} = \int_x^\infty p\,dx = \frac{1}{2} - \frac{1}{2}\,\frac{2}{\sqrt{\pi}}\int_0^{\frac{x}{2\sqrt{\pi}k\sqrt{t}}} e^{-\lambda^2}\,d\lambda,$$

pour que le cours x soit atteint ou dépassé à l'époque t, croit constamment avec le temps. Si t était infini, elle serait égale à $\frac{1}{2}$, résultat évident.

La probabilité

$$\int_{x_1}^{x_2} p\,dx = \frac{1}{\sqrt{\pi}}\int_{\frac{x_1}{2\sqrt{\pi}k\sqrt{t}}}^{\frac{x_2}{2\sqrt{\pi}k\sqrt{t}}} e^{-\lambda^2}\,d\lambda$$

pour que le cours se trouve compris à l'époque t, dans l'intervalle fini x_2, x_1, est nulle pour $t = 0$ et pour $t = \infty$. Elle est maxima lorsque

$$t = \frac{1}{4\,\pi\,k^2}\,\frac{x_2^2 - x_1^2}{\log\dfrac{x_2}{x_1}}.$$

Si nous supposons l'intervalle x_2, x_1 très petit, nous retrouvons pour époque de la probabilité maxima

$$t = \frac{x^2}{2\,\pi\,k^2}.$$

Écart probable. — Nous appellerons ainsi l'intervalle $\pm\,\alpha$ tel que, au bout du temps t, le cours ait autant de chances de rester compris dans cet intervalle que de chances de le dépasser.

La quantité α se détermine par l'équation

$$\int_0^\alpha p\,dx = \frac{1}{4}$$

ou

$$\Theta\left(\frac{\alpha}{2\,k\sqrt{\pi}\sqrt{t}}\right) = \frac{1}{2},$$

c'est-à-dire

$$\alpha = 2 \times 0,4769\, k \sqrt{\pi}\, \sqrt{t} = 1,688\, k \sqrt{t};$$

cet intervalle est proportionnel à la racine carrée du temps.

Plus généralement, considérons l'intervalle $\pm \beta$ tel que la probabilité pour que, à l'époque t, le cours soit compris dans cet intervalle soit égale à u, nous aurons

$$\int_0^\beta p\, dx = \frac{u}{2}$$

ou

$$\Theta\left(\frac{\beta}{2\, k \sqrt{\pi}\, \sqrt{t}}\right) = u.$$

Nous voyons que cet intervalle est proportionnel à la racine carrée du temps.

Rayonnement de la probabilité. — Je vais chercher directement l'expression de la probabilité $\mathcal{P}$ pour que le cours x soit atteint ou dépassé à l'époque t. Nous avons vu précédemment qu'en divisant le temps en intervalles très petits Δt, on pouvait considérer, pendant un intervalle Δt, le cours comme variant de la quantité fixe et très petite Δx.

Je suppose que, à l'époque t, les cours $x_{n-2}, x_{n-1}, x_n, x_{n+1}, x_{n+2}, \ldots$ différant entre eux de la quantité Δx, aient pour probabilités respectives : $p_{n-2}, p_{n-1}, p_n, p_{n+1}, p_{n+2}, \ldots$. De la connaissance de la distribution des probabilités à l'époque t, on déduit aisément la distribution des probabilités à l'époque $t + \Delta t$. Supposons, par exemple, que le cours x_n soit coté à l'époque t; à l'époque $t + \Delta t$ seront cotés les cours x_{n+1} ou x_{n-1}. La probabilité p_n, pour que le cours x_n soit coté à l'époque t, se décomposera en deux probabilités à l'époque $t + \Delta t$; le cours x_{n-1} aura de ce fait pour probabilité $\frac{p_n}{2}$, et le cours x_{n+1} aura du même fait pour probabilité $\frac{p_n}{2}$.

Si le cours x_{n-1} est coté à l'époque $t + \Delta t$, c'est que, à l'époque t, les cours x_{n-2} ou x_n ont été cotés; la probabilité du cours x_{n-1} à l'époque $t + \Delta t$ est donc $\frac{p_{n-2}+p_n}{2}$; celle du cours x_n est, à la même époque, $\frac{p_{n-1}+p_{n+1}}{2}$, celle du cours x_{n+1} est $\frac{p_n+p_{n+2}}{2}$, etc.

Pendant le temps Δt, le cours x_n a, en quelque sorte, émis vers le cours x_{n+1} la probabilité $\frac{p_n}{2}$; le cours x_{n+1} a, émis vers le cours x_n, la probabilité $\frac{p_{n+1}}{2}$. Si p_n est plus grand que p_{n+1}, l'échange de probabilité est $\frac{p_n - p_{n+1}}{2}$ de x_n vers x_{n+1}.

On peut donc dire :

Chaque cours x rayonne pendant l'élément de temps vers le cours voisin une quantité de probabilité proportionnelle à la différence de leurs probabilités.

Je dis proportionnelle, car on doit tenir compte du rapport de Δx à Δt.

La loi qui précède peut, par analogie avec certaines théories physiques, être appelée la *loi du rayonnement* ou de diffusion de la probabilité.

Je considère la probabilité $\mathcal{P}$ pour que le cours x se trouve à l'époque t dans l'intervalle x, ∞ et j'évalue l'accroissement de cette probabilité pendant le temps Δt.

Soit p la probabilité du cours x à l'époque t, $p = -\frac{d\mathcal{P}}{dx}$. Évaluons la probabilité qui, pendant le temps Δt, passe, en quelque sorte, à travers le cours x; c'est, d'après ce qui vient d'être dit,

$$\frac{1}{c^2}\left(p - \frac{dp}{dx} - p\right)\Delta t = -\frac{1}{c^2}\frac{dp}{dx}\Delta t = \frac{1}{c^2}\frac{d^2\mathcal{P}}{dx^2}\Delta t,$$

c désignant une constante.

Cet accroissement de probabilité a aussi pour expression $\frac{d\mathcal{P}}{dt}\Delta t$. On a donc

$$c^2\frac{\partial\mathcal{P}}{\partial t} - \frac{\partial^2\mathcal{P}}{\partial x^2} = 0.$$

C'est une équation de Fourier.

La théorie qui précède suppose les variations de cours discontinues; on peut arriver à l'équation de Fourier sans faire cette hypothèse, en remarquant que dans un intervalle de temps très petit Δt, le cours varie d'une façon continue, mais d'une quantité très petite inférieure à ε, par exemple.

Nous désignerons par ϖ la probabilité correspondant à p et relative à Δt.

D'après notre hypothèse, le cours ne pourra varier qu'à l'intérieur des limites $\pm \varepsilon$ dans le temps Δt et l'on aura par suite

$$\int_{-\varepsilon}^{+\varepsilon} \varpi \, dx = 1.$$

Le cours peut être $x - m$ à l'époque t; m étant positif et plus petit que ε; la probabilité de cette éventualité est p_{x-m}.

La probabilité pour que le cours x soit dépassé à l'époque $t + \Delta t$, ayant été égal à $x - m$ à l'époque t, aura pour valeur, en vertu du principe de la probabilité composée,

$$p_{x-m} \int_{\varepsilon-m}^{\varepsilon} \varpi \, dx.$$

Le cours peut être $x + m$, à l'époque t; la probabilité de cette éventualité est p_{x+m}.

La probabilité pour que le cours soit inférieur à x à l'époque $t + \Delta t$, ayant été égal à $x + m$ à l'époque t, aura pour valeur, en vertu du principe précédemment invoqué,

$$p_{x+m} \int_{\varepsilon-m}^{\varepsilon} \varpi \, dx.$$

L'accroissement de la probabilité $\mathcal{P}$, dans l'intervalle de temps Δt, sera égal à la somme des expressions telles que

$$(p_{x-m} - p_{x+m}) \int_{\varepsilon-m}^{\varepsilon} \varpi \, dx$$

pour toutes les valeurs de m depuis zéro jusqu'à ε.

Développons les expressions de p_{x-m} et p_{x+m} en négligeant les termes qui contiennent m^2, nous aurons

$$p_{x-m} = p_x - m \frac{dp_x}{dx},$$

$$p_{x+m} + p_x + m \frac{dp_x}{dx}.$$

L'expression ci-dessus devient alors

$$-\frac{dp}{dx}\int_{\varepsilon-m}^{\varepsilon} 2\,m\varpi\,dx.$$

L'accroissement cherché a donc pour valeur

$$-\frac{dp}{dx}\int_{0}^{\varepsilon}\int_{\varepsilon-m}^{\varepsilon} 2\,m\varpi\,dx\,dm.$$

L'intégrale ne dépend pas de x, de t ou de p, c'est une constante. L'accroissement de la probabilité $\mathfrak{P}$ a donc bien pour expression

$$\frac{1}{c^2}\frac{dp}{dx}.$$

L'équation de Fourier a pour intégrale

$$\mathfrak{P}=\int_{0}^{\infty} f\left(t-\frac{c^2 x^2}{2\alpha^2}\right)e^{-\frac{\alpha^2}{2}}d\alpha.$$

La fonction arbitraire f se détermine par les considérations suivantes :

On doit avoir $\mathfrak{P}=\frac{1}{2}$ si $x=0$, t ayant une valeur positive quelconque ; et $\mathfrak{P}=0$ lorsque t est négatif.

En posant $x=0$ dans l'intégrale ci-dessus, on a

$$\mathfrak{P}=f(t)\int_{0}^{\infty} e^{-\frac{\alpha^2}{2}}d\alpha=\frac{\sqrt{\pi}}{\sqrt{2}}f(t),$$

c'est-à-dire

$$f(t)=\frac{1}{\sqrt{2}\sqrt{\pi}} \quad \text{pour } t>0,$$
$$f(t)=0 \qquad \text{pour } t<0.$$

Cette dernière égalité nous montre que l'intégrale $\mathfrak{P}$ aura ses éléments nuls tant que $t-\frac{c^2 x^2}{2\alpha^2}$ sera plus petit que zéro, c'est-à-dire tant que α sera plus petit que $\frac{cx}{\sqrt{2}\sqrt{t}}$; on doit donc prendre pour limite inférieure dans l'intégrale $\mathcal{P}$, la quantité $\dfrac{cx}{\sqrt{2}\sqrt{t}}$ et l'on a

$$\mathcal{P} = \frac{1}{\sqrt{2}\sqrt{\pi}} \int_{\frac{cx}{\sqrt{2}\sqrt{t}}}^{\infty} e^{-\frac{\alpha^2}{2}} d\alpha = \frac{1}{\sqrt{\pi}} \int_{\frac{cx}{2\sqrt{t}}}^{\infty} e^{-\lambda^2} d\lambda,$$

ou, en remplaçant $\displaystyle\int_{\frac{cx}{2\sqrt{t}}}^{\infty}$ par $\displaystyle\int_{0}^{\infty} - \int_{0}^{\frac{cx}{2\sqrt{t}}}$,

$$\mathcal{P} = \frac{1}{2} - \frac{1}{2}\frac{2}{\sqrt{\pi}} \int_{0}^{\frac{cx}{2\sqrt{t}}} e^{-\lambda^2} d\lambda,$$

formule précédemment trouvée.

Loi des écarts de primes. — Pour connaître la loi qui régit le rapport de l'importance des primes et leurs écarts, nous appliquerons à l'acheteur de prime le principe de l'espérance mathématique :

L'espérance mathématique de l'acheteur de prime est nulle.

Prenons pour origine le cours vrai du ferme (*fig.* 5).

Soit p la probabilité au cours $\pm x$, c'est-à-dire dans le cas actuel la probabilité pour que la réponse des primes ait lieu au cours $\pm x$.

Soit $m + h$ l'écart vrai de la prime dont h.

Exprimons que l'espérance mathématique totale est nulle.

Fig. 5.

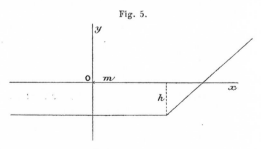

Nous évaluerons cette espérance :
1° Pour les cours compris entre $-\infty$ et m,
2°   »   » m et $m + h$,
3°   »   » $m + h$ et $+\infty$.

Ann. de l'Éc. Normale. 3ᵉ Série. Tome XVII. — FÉVRIER 1900.   7

145

1° Pour tous les cours compris entre $+\infty$ et m, la prime est abandonnée, c'est-à-dire que l'acheteur subit une perte h. Son espérance mathématique pour un cours compris dans l'intervalle donné est donc $-ph$ et pour tout l'intervalle

$$-h\int_{-\infty}^{m} p\,dx.$$

2° Pour un cours x compris entre m et $m+h$, la perte de l'acheteur sera $m+h-x$; l'espérance mathématique correspondante sera $-p(m+h-x)$ et pour l'intervalle entier

$$-\int_{m}^{m+h} p(m+h-x)\,dx.$$

3° Pour un cours x compris entre $m+h$ et ∞, le bénéfice de l'acheteur sera $x-m-h$; l'espérance mathématique correspondante sera $p(x-m-h)$ et pour tout l'intervalle

$$\int_{m+h}^{\infty} p(x-m-h)\,dx.$$

Le principe de l'espérance totale donnera donc

$$\int_{m+h}^{\infty} p(x-m-h)\,dx -\int_{m}^{m+h} p(m+h-x)\,dx - h\int_{-\infty}^{m} p\,dx = 0$$

ou, en faisant les réductions,

$$h + m\int_{m}^{\infty} p\,dx = \int_{m}^{\infty} px\,dx.$$

Telle est l'équation aux intégrales définies qui établit une relation entre les probabilités, les écarts de prime et leur importance.

Dans le cas où le pied de la prime tomberait du côté des x négatifs, comme le montre la *fig.* 6, m serait négatif et l'on arriverait à la relation

$$\frac{2h+m}{2} + m\int_{0}^{-m} p\,dx = \int_{-m}^{\infty} px\,dx.$$

Par suite de la symétrie des probabilités, la fonction p devant être paire, il en résulte que les deux équations ci-dessus n'en forment qu'une.

Fig. 6.

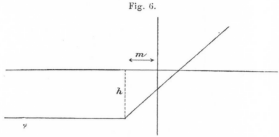

En différentiant, on obtient l'équation différentielle des écarts de prime

$$\frac{d^2 h}{dm^2} = p_m,$$

p_m étant l'expression de la probabilité dans laquelle on a remplacé x par m.

Prime simple. — Le cas le plus simple des équations ci-dessus est celui où $m = 0$, c'est-à-dire celui où l'importance de la prime est égale à son écart. On appelle *prime simple* cette sorte de prime, la seule que l'on traite dans la spéculation sur les marchandises.

Les équations ci-dessus deviennent, en posant $m = 0$ et en désignant par a la valeur de la prime simple,

$$a = \int_0^\infty p x\, dx = \int_0^\infty \frac{x}{2\pi k \sqrt{t}} e^{-\frac{x^2}{4\pi k^2 t}}\, dx = k \sqrt{t}.$$

L'égalité $a = \int_0^\infty p x\, dx$ montre que la prime simple est égale à l'espérance positive de l'acheteur ferme; ce fait est évident, puisque le preneur de prime verse la somme a au donneur pour jouir des avantages de l'acheteur ferme, c'est-à-dire pour avoir son espérance positive sans encourir ses risques.

De la formule

$$a = \int_0^\infty p x\, dx = k \sqrt{t},$$

147

nous déduisons le principe suivant, un des plus importants de notre étude :

La valeur de la prime simple doit être proportionnelle à la racine carrée du temps.

Nous avons vu précédemment que l'écart probable était donné par la formule

$$\alpha = 1,688\,k\sqrt{t} = 1,688\,a.$$

L'écart probable s'obtient donc en multipliant la prime moyenne par le nombre constant 1,688; il est donc très facile à calculer quand il s'agit de spéculations sur les marchandises puisque, dans ce cas, la quantité a est connue.

La formule suivante donne l'expression de la probabilité en fonction de a

$$p = \frac{1}{2\pi a}\, e^{-\frac{x^2}{4\pi a^2}}.$$

La probabilité dans un intervalle donné aura pour expression l'intégrale

$$\frac{1}{2\pi a}\int_0^u e^{-\frac{x^2}{4\pi a^2}}\,dx$$

ou

$$\frac{1}{2\pi a}\left(u - \frac{u^3}{12\pi a^2} + \frac{u^5}{160\pi^2 a^4} - \frac{u^7}{2678\pi^3 a^6} + \dots\right).$$

Cette probabilité est indépendante de a et, par suite, du temps, si u, au lieu d'être un nombre donné, est un paramètre de la forme $u = ba$; par exemple, si $u = a$,

$$\int_0^a p\,dx = \frac{1}{2\pi} - \frac{1}{24\pi^2} + \frac{1}{320\pi^3} - \dots = 0,155.$$

L'intégrale $\int_a^\infty p\,dx$ représente la probabilité de réussite du preneur de prime simple. Or

$$\int_a^\infty p\,dx = \frac{1}{2} - \int_0^a p\,dx = 0,345.$$

Donc :

La probabilité de réussite du preneur de prime simple est indépendante de l'époque de l'échéance; elle a pour valeur

$$0,345.$$

L'espérance positive de la prime simple a pour expression

$$\int_a^\infty p(x-a)\,dx = 0,58\,a.$$

Double prime. — Le *stellage* ou *double prime* est formé de l'achat simultané d'une prime à la hausse et à la baisse (primes simples). Il est facile de voir que le donneur de stellage est en bénéfice dans l'intervalle $-2a$, $+2a$; sa probabilité de réussite est donc

$$2\int_0^{2a} p\,dx = \frac{2}{\pi} - \frac{2}{3\pi^2} + \frac{2}{10\pi^3} - \ldots = 0,56.$$

La probabilité du preneur de stellage est $0,44$.

Espérance positive du stellage

$$2\int_{2a}^\infty p(x-2a)\,dx = 0,55\,a.$$

Coefficient d'instabilité. — Le coefficient k, précédemment introduit, est le *coefficient d'instabilité* ou de nervosité de la valeur, c'est lui qui mesure son état statique. Sa tension indique un état d'inquiétude; sa faiblesse, au contraire, est l'indice d'un état de calme.

Ce coefficient est donné directement dans la spéculation sur les marchandises par la formule

$$a = k\sqrt{t},$$

mais dans la spéculation sur les valeurs on ne peut le calculer que par approximation, comme nous allons le voir.

Série des écarts de prime. — L'équation aux intégrales définies des écarts de prime n'est pas exprimable en termes finis quand la quantité m, différence entre l'écart de la prime et son importance h, n'est pas nulle.

Cette équation conduit à la série

$$h - a + \frac{m}{2} - \frac{m^2}{4\pi a} + \frac{m^4}{96\pi^2 a^3} - \frac{m^6}{1920\pi^3 a^3} + \ldots = 0.$$

Cette relation, dans laquelle la quantité a désigne l'importance de la prime simple, permet de calculer la valeur de a quand on connait celle de m, ou inversement.

Loi approximative des écarts de prime. — La série qui précède peut s'écrire

$$h = a - f(m).$$

Considérons le produit de la prime h par son écart $(m + h)$:

$$h(m + h) = [a - f(m)][m + a - f(m)];$$

dérivons-en m, nous aurons

$$\frac{d}{dm}[h(m + h)] = f'(m)[m + a - f(m)] + [a - f(m)][1 - f'(m)].$$

Si nous posons $m = 0$, d'où $f(m) = 0$, $f'(m) = \frac{1}{2}$, cette dérivée s'annule ; nous devons en conclure que :

Le produit d'une prime par son écart est maximum quand les deux facteurs de ce produit sont égaux : c'est le cas de la prime simple.

Dans les environs de son maximum, le produit en question doit peu varier. C'est ce qui permet souvent de déterminer approximativement a par la formule

$$h(m + h) = a^2,$$

elle donne pour a une valeur trop faible.

En ne considérant que les trois premiers termes de la série, on obtient

$$h(h + m) = a^2 - \frac{m^2}{4},$$

qui donne pour a une valeur trop forte.

Dans la plupart des cas, en prenant les quatre premiers termes de la série, on obtiendra une approximation très suffisante; on aura ainsi

$$a = \frac{\pi(2h + m) \pm \sqrt{\pi^2(2h + m)^2 - 4\pi m^2}}{4\pi}.$$

Avec cette même approximation on aura pour valeur de m en fonction de a

$$m = \pi a \pm \sqrt{\pi^2 a^2 - 4\pi a(a - h)}.$$

Admettons pour un instant la formule simplifiée

$$h(m + h) = a^2 = k^2 t.$$

Dans la spéculation sur les valeurs les primes à la hausse ont une importance h constante, l'écart $m + h$ est donc proportionnel au temps.

L'écart des primes à la hausse, dans la spéculation sur les valeurs, est sensiblement proportionnel à la durée de leur échéance et au carré de l'instabilité.

Les primes à la baisse, sur les valeurs (c'est-à-dire, la vente ferme contre achat à prime) ont un écart h constant et une importance $m + h$ variable. Donc :

L'importance des primes à la baisse, dans la spéculation sur les valeurs, est sensiblement proportionnelle à la durée de leur échéance et au carré de l'instabilité.

Les deux lois qui précèdent ne sont qu'approchées.

Options. — Appliquons le principe de l'espérance mathématique à l'achat d'une option d'ordre n traitée à l'écart r.

L'option d'ordre n peut être considérée comme se composant de deux opérations :

1° Un achat ferme d'une unité au cours r;

2° Un achat ferme de $(n - 1)$ unités au cours r, cet achat n'étant à considérer que dans l'intervalle r, ∞.

La première opération a pour espérance mathématique $-r$, la seconde a pour espérance

$$(n-1)\int_r^\infty p(x-r)dx,$$

On doit donc avoir

$$r=(n-1)\int_r^\infty p(x-r)dx$$

ou, en remplaçant p par sa valeur,

$$p=\frac{1}{2\pi a}e^{-\frac{x}{4\pi a^2}},$$

et, en développant en série,

$$2\pi a^2-\pi a\frac{n+1}{n-1}r+\frac{r^2}{2}-\frac{r^4}{48\pi a^2}+\ldots=0,$$

En ne conservant que les trois premiers termes, on obtient

$$r=a\left[\frac{n+1}{n-1}\pi-\sqrt{\left(\frac{n+1}{n-1}\pi\right)^2-4\pi}\right].$$

Si $n=2$,

$$r=0,68\,a.$$

L'écart de l'option du double doit être environ les deux tiers de la valeur de la prime simple.

Si $n=3$,

$$r=1,096\,a.$$

L'écart de l'option du triple doit être supérieur de un dixième environ à la valeur de la prime simple.

Nous venons de voir que les écarts des options sont approximativement proportionnels à la quantité a.

Il en résulte que la probabilité de réussite de ces opérations est indépendante de la durée de l'échéance.

La probabilité de réussite de l'option du double est 0,394, *l'opération réussit quatre fois sur dix.*

La probabilité de l'option du triple est 0,33, *l'opération réussit une fois sur trois.*

L'espérance positive de l'option d'ordre n est

$$n \int_r^\infty p(x-r)\, dx,$$

et comme

$$\frac{r}{n-1} = \int_r^\infty p(x-r)\, dx,$$

l'espérance cherchée a pour valeur $\dfrac{n}{n-1}\, r$, c'est-à-dire 1,36a pour l'option du double et 1,64a pour l'option du triple.

En vendant ferme et en achetant simultanément une option du double, on obtient une prime dont l'importance est $r = 0,68a$ et dont l'écart est le double de r.

La probabilité de réussite de l'opération est 0,30.

Par analogie avec les opérations à prime, nous appellerons *option-stellage* d'ordre n, l'opération résultant de deux options d'ordre n, à la hausse et à la baisse.

L'option stellage du second ordre est une opération fort curieuse : entre les cours $\pm r$ la perte est constante et égale à $2r$. La perte diminue ensuite progressivement jusqu'aux cours $\pm 3r$, où elle s'annule.

Il y a bénéfice en dehors de l'intervalle $\pm 3r$.

La probabilité est 0,42.

OPÉRATIONS FERMES.

Maintenant que nous avons achevé l'étude générale des probabilités nous allons l'appliquer à la recherche des probabilités des principales opérations de bourse en commençant par les plus simples, les opérations fermes et les opérations à prime, et nous terminerons par l'étude des combinaisons de ces opérations.

La théorie de la spéculation sur les marchandises, beaucoup plus simple que celle des valeurs, a déjà été traitée ; nous avons, en effet, calculé la probabilité et l'espérance des primes simples des stellages et des options.

La théorie des opérations de bourse dépend de deux coefficients : *b* et *k*.

Leur valeur, à un instant donné, peut se déduire facilement de l'écart du terme au comptant et de l'écart d'une prime quelconque.

Dans l'étude qui va suivre, nous ne nous occuperons que de la rente 3 %, qui est une des valeurs sur laquelle on traite régulièrement des primes.

Nous prendrons pour valeurs de *b* et *k* leurs valeurs moyennes pour les cinq dernières années (1894 à 1898), c'est-à-dire

$$b = 0,264,$$
$$k = 5$$

(le temps est exprimé en jours et l'unité de variation est le centime).

Nous entendrons par valeurs *calculées* celles qui sont déduites des formules de la théorie avec les valeurs ci-dessus données aux constantes *b* et *k*.

Les valeurs *observées* sont celles que l'on déduit directement de la compilation des cotes durant ce même espace de temps de 1894 à 1898 ([1]).

Dans les Chapitres qui vont suivre nous aurons constamment à connaître les valeurs moyennes de la quantité *a* à différentes époques : la formule

$$a = 5\sqrt{t}$$

donne

Pour 45 jours...................... $a = 33,54$
»   30   »   .... $a = 27,38$
»   20   »   .... $a = 22,36$
»   10   »   .... $a = 16,13$

Pour un jour, il semble que l'on devrait avoir $a = 5$; mais dans tous les calculs de probabilités où il s'agit de moyennes on ne peut poser $t = 1$ pour un jour.

([1]) Toutes les observations sont extraites de la *Cote de la Bourse et de la Banque.*

En effet, il y a 365 jours dans l'année, mais seulement 307 jours de bourse. Le *jour moyen* de la bourse est donc $t = \frac{365}{307}$; il donne

$$a = 5,45.$$

On peut faire la même remarque pour le coefficient b. Dans tous les calculs relatifs à un jour de bourse on doit remplacer b par $b_1 = \frac{365}{307} b = 0,313.$

Écart probable. — Cherchons l'intervalle de cours $(-\alpha, +\alpha)$ tel que, au bout d'un mois, la rente ait autant de chances de se trouver dans cet intervalle que de chances de se trouver en dehors.

On devra avoir

$$\int_0^\alpha p\, dx = \frac{1}{4},$$

d'où

$$\alpha = \pm 46.$$

Pendant les 60 derniers mois, 33 fois la variation a été circonscrite entre ces limites et 27 fois elle les a dépassées.

On peut chercher de même l'intervalle relatif à un jour; on a ainsi

$$\alpha = \pm 9.$$

Sur 1452 observations, 815 fois la variation a été inférieure à 9^c.

Dans la question qui précède, nous avons supposé que le cours coté se confondait avec le cours vrai; dans ces conditions, la probabilité et l'espérance mathématique de l'acheteur et du vendeur sont les mêmes. En réalité, le cours coté est inférieur au cours vrai de la quantité nb, si n est le nombre de jours séparant de l'échéance.

L'écart probable de 46^c de part et d'autre du cours vrai correspond à l'intervalle compris entre 54^c en hausse au-dessus du cours coté et 38^c en baisse au-dessous de ce cours.

Formule de la probabilité dans le cas général. — Pour trouver la probabilité de la hausse pour une période de n jours, il faut connaître l'écart *nb* du cours vrai au cours coté; la probabilité est alors égale à

$$\int_{-nb}^{\infty} p\, dx.$$

La probabilité de la baisse sera égale à l'unité diminuée de la probabilité de la hausse.

Probabilité de l'achat au comptant. — Cherchons la probabilité de réussite d'un achat au comptant destiné à être revendu dans 3o jours.

On doit remplacer dans la formule précédente la quantité *nb* par 25.

La probabilité est alors égale à 0,64 :

L'opération a deux chances sur trois de réussir.

Si l'on veut avoir la probabilité pour un an, on doit remplacer la quantité *nb* par 3oo. La formule $a = k\sqrt{t}$ donne

$$a = 95,5.$$

On trouve que la probabilité est

$$0,89.$$

Neuf fois sur dix un achat de rente au comptant produit un bénéfice au bout d'un an.

Probabilité de l'achat ferme. — Cherchons la probabilité de réussite d'un achat ferme effectué au début du mois.

On a
$$nb = 7,91, \qquad a = 27,38.$$

On en déduit que :

        La probabilité de la hausse est.............. 0,55
           »   baisse   » 0,45

La probabilité de l'achat croît avec le temps; pour un an, on a

$$n = 365, \qquad nb = 96,36, \qquad a = 95,5.$$

La probabilité a alors pour valeur 0,65.

Quand on effectue un achat ferme pour le revendre au bout d'un an, on a deux chances sur trois de réussir.

156

Il est évident que si le report mensuel était de 25ᶜ la probabilité de l'achat serait 0,50.

Avantage mathématique des opérations fermes. — Il me paraît indispensable, comme je l'ai déjà fait remarquer, d'étudier l'avantage mathématique d'un jeu dès qu'il n'est pas équitable, et c'est le cas des opérations fermes.

Si nous supposons $b = 0$, l'espérance mathématique de l'achat ferme est $a - a = 0$. L'avantage de l'opération est $\frac{a}{2a} = \frac{1}{2}$ comme d'ailleurs dans tout jeu équitable.

Cherchons l'avantage mathématique d'un achat ferme de n jours en supposant $b > 0$. L'acheteur aura, pendant cette période, touché la somme nb provenant de la différence entre les coupons et les reports, et son espérance sera $a - a + nb$; son avantage mathématique sera donc

$$\frac{a + nb}{2a + nb}.$$

L'avantage du vendeur serait

$$\frac{a}{2a + nb}.$$

Occupons-nous spécialement du cas de l'acheteur.

Quand $b > 0$ son avantage mathématique croît de plus en plus avec n; il est constamment supérieur à la probabilité.

Pour un mois, l'avantage de l'acheteur est 0,563 et sa probabilité 0,55. Pour un an, son avantage est 0,667 et sa probabilité 0,65.

On peut donc dire que :

L'avantage d'une opération ferme est à peu près égal à sa probabilité.

OPÉRATIONS A PRIME.

Écart des primes. — Connaissant la valeur de a (pour une époque donnée, on calcule facilement l'écart vrai par la formule

$$m = \pi a \pm \sqrt{\pi^2 a^2 - 4\pi a(a - h)}.$$

Connaissant l'écart vrai on obtient l'écart coté en ajoutant la quantité nb à l'écart vrai ; n est le nombre de jours qui séparent de la réponse.

Dans le cas d'une prime fin prochain, on ajoute la quantité $[25 + (n - 30)b]$.

On arrive ainsi aux résultats suivants :

Primes dont 50.

	Écart coté	
	calculé.	observé.
A 45 jours..............	50,01	52,62
30 »	20,69	21,22
20 »	13,23	14,71

Primes dont 25.

	Écart coté	
	calculé.	observé.
A 45 jours..............	72,70	72,80
30 »	37,78	37,84
20 »	25,17	27,39
10 »	12,24	17,40

Primes dont 10.

	Écart coté	
	calculé.	observé.
A 30 jours..............	66,19	60,93
20 »	48,62	46,43
10 »	26,91	32,89

Dans le cas de la prime dont 5e pour le lendemain nous avons

$$h = 5, \qquad a = 5,45$$

d'où

$$m = 0,81;$$

l'écart vrai est donc 5,81 ; en y ajoutant $b_1 = \dfrac{365}{307} b = 0,31$ on obtient l'écart calculé 6,12.

La moyenne des cinq dernières années donne 7,36.

Les chiffres observés et calculés concordent dans leur ensemble, mais ils présentent certaines divergences qu'il est indispensable d'expliquer.

Ainsi l'écart observé de la prime dont 10 à 30 jours est trop faible; il est facile d'en comprendre la raison : Dans les périodes très mouvementées, alors que la prime dont 10 serait à un très fort écart, on ne cote pas cette prime; la moyenne observée se trouve donc diminuée de ce fait.

D'autre part, il n'est pas niable que le marché ait eu, pendant plusieurs années, une tendance à coter à de trop forts écarts les primes à courtes échéances; il se rend d'autant moins compte de la juste proportion des écarts que ceux-ci sont plus petits et que l'échéance est plus proche.

Il faut cependant ajouter qu'il semble s'être aperçu de son erreur, car en 1898 il a paru exagérer dans le sens inverse.

Probabilité de levée des primes. — Pour qu'une prime soit levée, il faut que le cours de la réponse des primes soit supérieur au cours du pied de la prime; la probabilité de levée est donc exprimée par l'intégrale

$$\int_\varepsilon^\infty p\,dx,$$

ε étant le cours vrai du pied de la prime.

Cette intégrale est facile à calculer, comme on l'a vu précédemment; elle conduit aux résultats suivants :

Probabilité de levée des primes dont 50.

	Calculée.	Observée.
A 45 jours...............	0,63	0,59
30 »	0,71	0,75
20 »	0,77	0,76

Probabilité de levée des primes dont 25.

	Calculée.	Observée.
A 45 jours...............	0,41	0,40
30 »	0,47	0,46
20 »	0,53	0,53
10 »	0,65	0,65

Probabilité de levée des primes dont 10.

	Calculée.	Observée.
A 30 jours.............	0,24	0,21
20 »	0,28	0,26
10 »	0,36	0,38

On peut dire que les primes /50 sont levées trois fois sur quatre, les primes /25 deux fois sur quatre et les primes /10 une fois sur quatre.

La probabilité de levée de la prime dont 5^c pour le lendemain est, d'après le calcul : 0,48; le résultat de 1456 observations donne 671 primes certainement levées et 76 dont la levée est douteuse; en comptant ces 76 dernières primes la probabilité serait 0,51, en ne les comptant pas elle serait 0,46, soit en moyenne 0,48 comme l'indique la théorie.

Probabilité de bénéfice des primes. — Pour qu'une prime donne du bénéfice à son acheteur, il faut que la réponse des primes se fasse à un cours supérieur à celui de la prime. La probabilité de bénéfice est donc exprimée par l'intégrale

$$\int_{\varepsilon_1}^{\infty} p\, dx,$$

ε_1 étant le cours de la prime.

Cette intégrale conduit aux résultats ci-après :

Probabilité de bénéfice des primes dont 50.

	Calculée.	Observée.
A 45 jours.............	0,40	0,39
30 »	0,43	0,41
20 »	0,44	0,40

Probabilité de bénéfice des primes dont 25.

	Calculée.	Observée.
A 45 jours.............	0,30	0,27
30 »	0,33	0,31
20 »	0,36	0,30
10 »	0,41	0,40

Probabilité de bénéfice des primes dont 10.

	Calculée.	Observée.
A 3o jours..............	0,20	0,16
20 »	0,22	0,18
10 »	0,27	0,25

On voit qu'entre les limites ordinaires de la pratique, la probabilité de réussite de l'achat d'une prime varie peu. L'achat /5o réussit quatre fois sur dix, l'achat /25 trois fois sur dix et l'achat /10 deux fois sur dix.

D'après le calcul, l'acheteur de prime dont 5ᶜ pour le lendemain a une probabilité de réussite de 0,34, l'observation de 1456 cotes montre que 410 primes auraient certainement donné des bénéfices et que 80 autres donnent un résultat douteux, la probabilité observée est donc 0,31.

OPÉRATIONS COMPLEXES.

Classification des opérations complexes. — Comme on traite du ferme et souvent jusqu'à trois primes pour la même échéance, on pourrait entreprendre en même temps des opérations triples et même quadruples.

Les opérations triples sortent déjà du nombre de celles que l'on peut considérer comme classiques, leur étude est très intéressante, mais trop longue pour pouvoir être exposée ici. Nous nous bornerons donc aux opérations doubles.

On peut les diviser en deux groupes suivant qu'elles contiennent ou non du ferme.

Les opérations contenant du ferme se composeront d'un achat ferme et d'une vente à prime, ou inversement.

Les opérations à prime contre prime consistent dans la vente d'une grosse prime suivie de l'achat d'une petite, ou inversement.

La proportion des achats et des ventes peut d'ailleurs varier à l'infini. Pour simplifier la question, nous n'étudierons que deux proportions très simples :

1° La seconde opération porte sur le même chiffre que la première.

2° Elle porte sur un chiffre double.

Pour fixer les idées, nous supposerons que l'on opère au début du mois et nous prendrons pour écarts vrais les écarts moyens depuis cinq ans : 12,78/5o, 29,87/25 et 58,28/10.

Nous remarquerons aussi que pour les opérations à un mois le cours vrai est plus élevé que le cours coté, de la quantité $7,91 = 3ob$.

Achat ferme contre vente à prime. — On achète en réalité du ferme au cours $— 3ob = — 7,91$ et l'on vend à prime /25 au cours $+ 29,87$.

Il est facile de représenter l'opération par une construction géométrique (*fig.* 7) : l'achat ferme est représenté par la droite AMB:

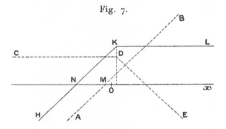

Fig. 7.

$MO = 3ob$. La vente à prime est représentée par la ligne brisée CDE, l'opération résultante sera représentée par la ligne brisée HNKL, l'abscisse du point N sera

$$— (25 + 3ob).$$

On voit que l'opération donne un bénéfice limité égal à l'écart coté de la prime; à la baisse, le risque est illimité.

La probabilité de réussite de l'opération est exprimée par l'intégrale

$$\int_{-25-30b}^{+\infty} p\,dx = 0,68.$$

Si l'on avait vendu une prime /5o la probabilité de réussite aurait été : 0,80.

Il est intéressant de connaître la probabilité dans le cas d'un report de 25^c ($b = o$).

Cette probabilité est 0,64 en vendant /25 et 0,76 en vendant /50.

Si l'on revend une prime sur un achat au comptant, la probabilité est 0,76 en revendant /25 et 0,86 en revendant /50.

Vente ferme contre achat à prime. — Cette opération est inverse de la précédente ; elle donne à la hausse une perte limitée et à la baisse un bénéfice illimité. C'est, par conséquent, une prime à la baisse, prime dont l'écart est constant et l'importance variable, à l'inverse des primes à la hausse.

Achat ferme contre vente du double à prime. — On achète ferme au cours vrai — 30b et l'on vend le double au cours 29,87/25.

La *fig.* 8 représente géométriquement l'opération ; elle montre que le risque est illimité à la hausse comme à la baisse.

Fig. 8.

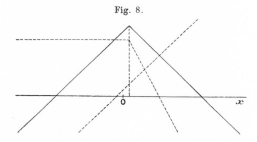

On gagne entre les cours — (50 + 30b) et 59,74 + 30b. La probabilité de réussite

$$\int p \, dx = 0,64.$$

En vendant /50 la probabilité serait 0,62 et en vendant /10 on aurait pour probabilité 0,62.

Si l'on avait acheté 2 unités ferme pour en vendre 3/50, la probabilité aurait été 0,66.

Vente ferme contre achat du double à prime. — C'est l'opération in-

verse de la précédente; elle donne des bénéfices dans le cas d'une forte
hausse et dans celui d'une forte baisse.

Sa probabilité est : 0,27.

Achat d'une grosse prime contre vente d'une petite. — Je suppose
qu'on ait fait simultanément les deux opérations suivantes :

$$\text{Acheté à}\dots\dots\dots\dots\dots\dots\dots\dots \quad 12,78/50$$
$$\text{Vendu à}\dots\dots\dots\dots\dots\dots\dots\dots \quad 29,87/25$$

Au-dessous du pied de la grosse prime $(-37,22)$, les deux primes
sont abandonnées et l'on perd 25^c.

A partir du cours $-37,22$ on est acheteur, et au cours de $-12,22$
l'opération est nulle. On gagne ensuite jusqu'à ce que le pied de la
prime $/25$, c'est-à-dire le cours $+4,87$ soit atteint.

Alors on est liquidé et l'on gagne l'écart. En baisse on perd donc
25^c, c'est le risque maximum; en hausse on gagne l'écart.

Le risque est limité, le bénéfice l'est également.

La *fig.* 9 représente géométriquement l'opération.

Fig. 9.

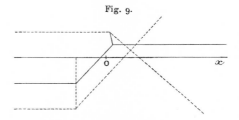

La probabilité de réussite est donnée par l'intégrale

$$\int_{-12,22}^{\infty} p\,dx = 0,59.$$

En achetant $/25$ pour vendre $/10$, la probabilité de réussite serait
$0,38$.

Vente d'une grosse prime contre achat d'une petite. — Cette opération,
qui est la contre-partie de la précédente, se discute sans difficulté; en

baisse on gagne la différence du montant des primes, en hausse on
perd leur écart.

Achat d'une grosse prime contre vente d'une petite en quantité double.
— Je suppose qu'on ait fait l'opération suivante :

Achat à............................... 12,78/50
Vente du double....................... 29,87/25

En forte baisse, les primes sont abandonnées, elles se compensent ;
c'est une opération *en blanc*. Au pied de la grosse prime, c'est-à-dire
au cours — 37,22, on devient acheteur et l'on gagne progressivement
jusqu'au pied de la petite (+ 4,87).

A ce moment, le bénéfice est maximum (42,09 centimes) et l'on
devient vendeur. On reperd progressivement le bénéfice et au cours
de 45,96 ce bénéfice est nul.

Au delà on perd proportionnellement à la hausse.

En résumé, l'opération donne un bénéfice limité, un risque nul à la
baisse et illimité à la hausse.

La *fig.* 10 représente géométriquement l'opération.

Fig. 10.

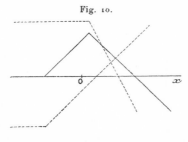

Probabilité de l'opération en blanc........... 0,30
     »   de bénéfice..................... 0,45
     »   de perte 0,25

Vente d'une grosse prime contre achat d'une petite en quantité double.
— La discussion et la représentation géométrique de cette opération,
inverse de la précédente, ne présentent aucune difficulté. Il est inutile
de nous y arrêter.

165

Classification pratique des opérations de bourse. — Au point de vue pratique, on peut diviser les opérations de bourse en quatre classes :

Les opérations à la hausse.

Les opérations à la baisse.

Les opérations en prévision d'un grand mouvement dans un sens quelconque.

Les opérations en prévision des petits mouvements.

Le Tableau suivant résume les principales opérations à la hausse :

	Probabilité moyenne.		
	$b = \dfrac{25}{30}$ (report nul).	$b = 0,26$ (report moyen).	$b = 0$ (report égal aux coupons).
Achat /10	0,20	0,20	0,20
Achat /25	0,33	0,33	0,33
Achat /25 C vente /10	0,38	0,38	0,38
Achat /50	0,43	0,43	0,43
Achat ferme	0,64	0,55	0,50
Achat /50 C vente /25	0,59	0,59	0,59
Achat ferme C vente /25	0,76	0,68	0,64
» » /50	0,86	0,80	0,76

Il suffit d'inverser ce Tableau pour obtenir l'échelle des opérations à la baisse.

PROBABILITÉ POUR QU'UN COURS SOIT ATTEINT DANS UN INTERVALLE DE TEMPS DONNÉ.

Cherchons la probabilité P pour qu'un cours donné c soit atteint ou dépassé dans un intervalle de temps t.

Supposons d'abord, pour simplifier, que le temps soit décomposé en deux unités, que t égale deux jours par exemple.

Soit x le cours coté le premier jour et soit y le cours du second jour relativement à celui du premier.

Pour que le cours c soit atteint ou dépassé, il faut que le premier jour le cours soit compris entre c et ∞ ou que, le second jour, il soit compris entre $c - x$ et ∞.

Dans la question actuelle, il faut distinguer quatre cas :

1er jour. x compris entre :			2e jour. y compris entre :		
$-\infty$	et	c	$-\infty$	et	$c - x$
$-\infty$	et	c	$c - x$	et	$+\infty$
c	et	∞	$-\infty$	et	$c - x$
c	et	∞	$c - x$	et	$+\infty$

Sur ces quatre cas, les trois derniers sont favorables.

La probabilité, pour que le cours se trouve compris dans l'intervalle dx le premier jour et dans l'intervalle dy le second jour, sera $p_x p_y\, dx\, dy$.

La probabilité P, étant par définition le rapport du nombre des cas favorables à celui des cas possibles, aura pour expression

$$P = \frac{\int_{-\infty}^{c}\int_{c-x}^{\infty} + \int_{c}^{\infty}\int_{-\infty}^{c-x} + \int_{c}^{\infty}\int_{c-x}^{\infty}}{\int_{-\infty}^{c}\int_{-\infty}^{c-x} + \int_{-\infty}^{c}\int_{c-x}^{\infty} + \int_{c}^{\infty}\int_{-\infty}^{c-x} + \int_{c}^{\infty}\int_{c-x}^{\infty}}$$

(l'élément est $p_x p_y\, dx\, dy$).

Les quatre intégrales du dénominateur représentent les quatre cas possibles ; les trois intégrales du numérateur représentent les trois cas favorables. On peut simplifier et écrire, le dénominateur étant égal à un,

$$P = \int_{-\infty}^{c}\int_{c-x}^{\infty} p_x p_y\, dx\, dy + \int_{c}^{\infty}\int_{-\infty}^{\infty} p_x p_y\, dx\, dy.$$

On pourrait appliquer le même raisonnement en supposant que l'on ait à considérer trois jours consécutifs, puis quatre, etc.

Cette méthode conduirait à des expressions de plus en plus compliquées, car le nombre des cas favorables irait sans cesse en augmentant. Il est beaucoup plus simple d'étudier la probabilité $1 - P$ pour que le cours c ne soit jamais atteint.

Il n'y a plus alors qu'un seul cas favorable quel que soit le nombre de jours, c'est celui où le cours n'est atteint à aucun des jours considérés.

167

La probabilité $1 - P$ a pour expression

$$1 - P = \int_{-\infty}^{c} \int_{-\infty}^{c-x_1} \int_{-\infty}^{c-x_1-x_2} \cdots \int_{-\infty}^{c-x_1\cdots-x_{n-1}} p_{x_1} \cdots p_{x_n}\, dx_1 \cdots dx_n,$$

x_1 est le cours du premier jour;

x_2 est le cours du second jour relativement à celui du premier;

x_3 est le cours relatif du troisième jour, etc.

La détermination de cette intégrale paraissant difficile, nous résoudrons la question en employant une méthode d'approximation.

On peut considérer le temps t comme divisé en petits intervalles Δt de telle sorte que $t = m\Delta t$. Pendant l'unité de temps Δt, le cours ne variera que de la quantité $\pm \Delta x$, écart moyen relatif à cette unité de temps.

Chacun des écarts $\pm \Delta x$ aura pour probabilité $\frac{1}{2}$,

Supposons que $c = n\Delta x$ et cherchons la probabilité pour que le cours c soit atteint précisément à l'époque t; c'est-à-dire pour que ce cours soit atteint à cette époque t, sans l'avoir jamais été antérieurement. Si, pendant les m unités de temps, le cours a varié de la quantité $n\Delta x$, c'est qu'il y a eu $\frac{m+n}{2}$ variations en hausse et $\frac{m-n}{2}$ variations en baisse.

La probabilité pour que, sur m variations, il y en ait eu $\frac{m+n}{2}$ favorables est

$$\frac{m!}{\dfrac{m-n}{2}!\ \dfrac{m+n}{2}!}\left(\frac{1}{2}\right)^m.$$

Ce n'est pas cette probabilité que nous cherchons, mais le produit de cette probabilité par le rapport du nombre des cas où le cours $n\Delta x$ est atteint à l'époque $m\Delta t$, ne l'ayant pas été précédemment, au nombre total des cas où il est atteint à l'époque $m\Delta t$.

Nous allons calculer ce rapport.

Pendant les m unités de temps que nous considérons, il y a eu $\frac{m+n}{2}$ variations en hausse et $\frac{m-n}{2}$ variations en baisse.

Nous pouvons représenter une des combinaisons donnant une hausse de $n\,\Delta x$ en m unités de temps par le symbole

$$\mathrm{B_1\,H_1\,H_2 \ldots B_{\frac{m-n}{2}} \ldots H_{\frac{m+n}{2}}},$$

$\mathrm{B_1}$ indique que, pendant la première unité de temps, il y a eu baisse; $\mathrm{H_1}$, qui vient ensuite, indique qu'il y a eu hausse pendant la seconde unité de temps, etc.

Pour qu'une combinaison soit favorable, il faut que, en la lisant de droite à gauche, le nombre des H soit constamment supérieur à celui des B. Nous sommes ramenés, comme on voit, au problème suivant :

Sur n lettres il y a $\dfrac{m+n}{2}$ lettres H et $\dfrac{m-n}{2}$ lettres B; quelle est la probabilité pour que, en écrivant ces lettres au hasard et en les lisant dans un sens déterminé, le nombre des H soit, durant toute la lecture, toujours supérieur à celui des B?

La solution de ce problème, présenté sous une forme un peu différente, a été donnée par M. André. La probabilité cherchée est égale à $\dfrac{n}{m}$.

La probabilité pour que le cours $n\,\Delta x$ soit atteint précisément au bout de m unités de temps est donc

$$\frac{n}{m} \; \frac{m!}{\frac{m-n}{2}! \; \frac{m+n}{2}!} \left(\frac{1}{2}\right)^m.$$

Cette formule est approximative; nous obtiendrons une expression plus exacte en remplaçant la quantité qui multiplie $\dfrac{n}{m}$ par la valeur exacte de la probabilité à l'époque t, c'est-à-dire par

$$\frac{\sqrt{2}}{\sqrt{m}\sqrt{\pi}}\, e^{-\frac{n^2}{\pi m}}.$$

La probabilité que nous cherchons est donc

$$\frac{n\sqrt{2}}{m\sqrt{m}\sqrt{\pi}}\, e^{-\frac{n^2}{\pi m}},$$

ou, en remplaçant n par $\dfrac{2\,c\sqrt{\pi}}{\sqrt{2}}$ et m par $8\,\pi\,k^2\,t$,

$$\frac{dt\,c\sqrt{2}}{2\sqrt{\pi}\,kt\sqrt{t}}\,e^{-\frac{c^2}{4\pi k^2 t}}.$$

Telle est l'expression de la probabilité pour que le cours c soit atteint à l'époque dt, ne l'ayant pas été antérieurement.

La probabilité pour que le cours c ne soit pas atteint avant l'époque t aura pour valeur

$$1 - P = A\int_{t}^{\infty}\frac{c\sqrt{2}}{2\sqrt{\pi}\,kt\sqrt{t}}\,e^{-\frac{c^2}{4\pi k^2 t}}\,dt.$$

J'ai multiplié l'intégrale par une constante à déterminer A, parce que le cours ne peut être atteint que si la quantité désignée par m est paire.

En posant

$$\lambda^2 = \frac{c^2}{4\,\pi\,k^2\,t},$$

on a

$$1 - P = 2\sqrt{2}\,A\int_{0}^{\frac{c}{2\sqrt{\pi}k\sqrt{t}}}e^{-\lambda^2}\,d\lambda.$$

Pour déterminer A, posons $c = \infty$, alors $P = 0$ et

$$1 = 2\sqrt{2}\,A\int_{0}^{\infty}e^{-\lambda^2}\,d\lambda = \sqrt{2}\sqrt{\pi}\,A;$$

donc

$$A = \frac{1}{\sqrt{2}\sqrt{\pi}},$$

alors

$$1 - P = \frac{2}{\sqrt{\pi}}\int_{0}^{\frac{c}{2\sqrt{\pi}k\sqrt{t}}}e^{-\lambda^2}\,d\lambda.$$

La probabilité, pour que le cours x soit atteint ou dépassé pendant l'in-

tervalle de temps t a donc pour expression

$$P = 1 - \frac{2}{\sqrt{\pi}} \int_0^{\frac{x}{2\sqrt{\pi}\,k\sqrt{t}}} e^{-\lambda^2}\, d\lambda.$$

La probabilité pour que le cours x soit atteint ou dépassé *à l'époque t* a pour expression, comme nous l'avons vu,

$$\mathcal{P} = \frac{1}{2} - \frac{1}{2}\frac{2}{\sqrt{\pi}} \int_0^{\frac{x}{2\sqrt{\pi}k\sqrt{t}}} e^{-\lambda^2}\, d\lambda.$$

On voit que $\mathcal{P}$ est la moitié de P.

La probabilité pour qu'un cours soit atteint ou dépassé à l'époque t est la moitié de la probabilité pour que ce cours soit atteint ou dépassé dans l'intervalle de temps t.

La démonstration directe de ce résultat est très simple : Le cours ne peut être dépassé à l'époque t sans l'avoir été antérieurement. La probabilité $\mathcal{P}$ est donc égale à la probabilité P, multipliée par la probabilité pour que, le cours étant coté à une époque antérieure à t, soit dépassé à l'époque t; c'est-à-dire, multipliée par $\frac{1}{2}$. On a donc

$$\mathcal{P} = \frac{P}{2}.$$

On peut remarquer que l'intégrale multiple qui exprime la probabilité $1 - P$ et qui semble réfractaire aux procédés ordinaires de calcul se trouve déterminée par un raisonnement très simple grâce au calcul des probabilités.

Applications. — Les Tables de la fonction Θ permettent de calculer très facilement la probabilité

$$P = 1 - \Theta\left(\frac{x}{2\sqrt{\pi}\,k\sqrt{t}}\right).$$

La formule

$$P = 1 - \frac{2}{\sqrt{\pi}} \int_0^{\frac{x}{2\sqrt{\pi}k\sqrt{t}}} e^{-\lambda^2} d\lambda$$

montre que la probabilité est constante, quand l'écart x est proportionnel à la racine carrée du temps ; c'est-à-dire, quand il a une expression de la forme $x = ma$. Nous allons étudier les probabilités correspondant à certains écarts intéressants.

Supposons d'abord que $x = a = k\sqrt{t}$; la probabilité P est alors égale à 0,69. Quand l'écart a est atteint, on peut, sans perte, revendre du ferme sur la prime simple a. Donc :

Il y a deux chances sur trois pour que l'on puisse, sans perte, revendre du ferme sur une prime simple.

Particularisons la question en l'appliquant à la rente 3 pour 100 ; sur une période de 60 mois, 38 fois, on a pu revendre à l'écart a ; ce qui correspond à une probabilité de 0,63.

Étudions maintenant le cas où $x = 2a$.
La formule précédente donne pour probabilité 0,43.
Quand l'écart $2a$ est atteint, on peut revendre sans perte du ferme sur une prime double ; ainsi :

Il y a quatre chances sur dix pour que l'on puisse, sans perte, revendre du ferme sur une prime double.

Sur une période de 60 liquidations, la rente 3 pour 100 a atteint 23 fois l'écart $2a$, ce qui donne pour probabilité 0,38.

L'écart $0,7a$ est celui de l'option du double ; la probabilité correspondante est 0,78.

On a trois chances sur quatre de pouvoir, sans perte, revendre du ferme sur une option du double.

L'option du triple doit se traiter à un écart $1,1a$ auquel correspond la probabilité 0,66.

On a deux chances sur trois de pouvoir, sans perte, revendre du ferme sur une option du triple.

Citons, enfin, comme écarts remarquables l'écart $1,7a$ qui correspond à une probabilité de $\frac{1}{2}$ et l'écart $2,9a$ qui correspond à une probabilité de $\frac{1}{4}$.

Espérance mathématique apparente. — L'espérance mathématique

$$\mathcal{E}_1 = \mathrm{P}\,x = x - \frac{2x}{\sqrt{\pi}}\int_0^{\frac{x}{2\sqrt{\pi k\sqrt{t}}}} e^{-\lambda^2}\,d\lambda$$

est une fonction de x et de t; différentions-la par rapport à x, nous aurons

$$\frac{\partial\mathcal{E}_1}{\partial x} = 1 - \frac{2}{\sqrt{\pi}}\int_0^{\frac{x}{2\sqrt{\pi k\sqrt{t}}}} e^{-\lambda^2}\,d\lambda - \frac{x\,e^{-\frac{x^2}{4\pi k^2 t}}}{\pi k\sqrt{t}}.$$

Si l'on considère une époque déterminée t, cette espérance sera maxima lorsque

$$\frac{\partial\mathcal{E}_1}{\partial x} = 0,$$

c'est-à-dire, quand $x = 2a$, environ.

Espérance totale apparente. — L'espérance totale correspondant au temps t sera l'intégrale

$$\int_0^\infty \mathrm{P}x\,dx.$$

Posons

$$f(a) = \int_0^\infty \left(x - \frac{2x}{\sqrt{\pi}}\int_0^{\frac{x}{2\sqrt{\pi}\,t}} e^{-\lambda^2}\,d\lambda \right) dx.$$

Différentions en a, nous aurons

$$f'(a) = \frac{1}{\pi a^2}\int_0^\infty x^2\,e^{-\frac{x^2}{4\pi a^2}}\,dx,$$

ou $f'(a) = 2\pi a$. On a donc

$$f(a) = \pi a^2 = \pi k^2 t.$$

L'espérance totale est proportionnelle au temps.

Époque de la plus grande probabilité. — La probabilité

$$P = 1 - \frac{2}{\sqrt{\pi}} \int_0^{\frac{x}{2\sqrt{\pi}k\sqrt{t}}} e^{-\lambda^2}\, d\lambda$$

est une fonction de x et de t.

L'étude de sa variation, en considérant x comme variable, ne présente aucune particularité; la fonction décroît constamment quand x croît.

Supposons maintenant que x soit constant et étudions la variation de la fonction en considérant t comme variable, nous aurons en différentiant

$$\frac{\partial P}{\partial t} = \frac{x e^{-\frac{x^2}{4\pi k^2 t}}}{2\pi t \sqrt{t}}.$$

Nous déterminerons l'époque de la probabilité maxima en annulant la dérivée

$$\frac{\partial^2 P}{\partial t^2} = \frac{x e^{-\frac{x^2}{4\pi k^2 t}}}{2\pi k t \sqrt{t}} \left(\frac{x^2}{4\pi k^2 t} - \frac{3}{2} \right);$$

on a alors

$$t = \frac{x^2}{6\pi k^2}.$$

Supposons, par exemple, que $x = k\sqrt{t_1}$, nous aurons $t = \frac{t_1}{6\pi}$.

L'époque la plus probable à laquelle on peut sans perte revendre du ferme sur une prime simple est située au dix-huitième de la durée de l'échéance.

Si nous supposons maintenant que $x = 2k\sqrt{t_1}$, nous obtenons $t = \frac{2t_1}{3\pi}$.

L'époque la plus probable à laquelle on peut sans perte revendre du ferme sur une prime double est située au cinquième de la durée de l'échéance.

La probabilité P correspondant à l'époque $t = \dfrac{x^2}{6\pi k^2}$ a pour valeur

$$1 - \Theta\left(\frac{\sqrt{6}}{2}\right) = 0,08.$$

Époque moyenne. — Lorsqu'un événement peut se produire à différentes époques, on appelle époque moyenne de l'arrivée de l'événement la somme des produits des probabilités correspondant aux époques données par leurs durées respectives.

La durée moyenne est égale à la somme des espérances de durée.

L'époque moyenne à laquelle le cours x sera dépassé est donc exprimée par l'intégrale

$$\int_0^\infty t\,\frac{d\mathrm{P}}{dt}\,dt = \int_0^\infty \frac{x}{2\pi k\sqrt{t}}\,e^{-\frac{x^2}{4\pi k^2 t}}\,dt;$$

en posant $\dfrac{x^2}{4\pi k^2 t} = y^2$, elle devient

$$\frac{x^2}{2\pi\sqrt{\pi}\,k^2}\int_0^\infty \frac{e^{-y^2}}{y^2}\,dy.$$

Cette intégrale est infinie.
L'époque moyenne est donc infinie.

Époque probable absolue. — Ce sera l'époque pour laquelle on aura $\mathrm{P} = \dfrac{1}{2}$ ou

$$\Theta\left(\frac{x}{2\sqrt{\pi}\,k\sqrt{t}}\right) = \frac{1}{2};$$

on en déduit

$$t = \frac{x^2}{2,89\,k^2}.$$

L'époque probable absolue varie, de même que l'époque la plus probable, proportionnellement au carré de la quantité x, et elle est environ six fois supérieure à l'époque la plus probable.

Époque probable relative — Il est intéressant de connaître, non seulement la probabilité pour qu'un cours x soit coté dans un intervalle de temps t, mais encore l'époque probable T à laquelle ce cours doit être atteint; cette époque est évidemment différente de celle dont nous venons de nous occuper.

L'intervalle de temps T sera tel qu'il y aura autant de chances pour que le cours soit atteint avant l'époque T que de chances pour qu'il soit coté dans la suite, c'est-à-dire dans l'intervalle de temps T,t.

T sera donné par la formule

$$\int_0^T \frac{\partial P}{\partial t} dt = \frac{1}{2} \int_0^t \frac{\partial P}{\partial t} dt$$

ou

$$1 - 2\Theta\left(\frac{x}{2\sqrt{\pi}\,k\sqrt{T}}\right) = -\Theta\left(\frac{x}{2\sqrt{\pi}\,k\sqrt{t}}\right).$$

Comme application, supposons que $x = k\sqrt{t}$; la formule donne T $= 0,18t$; donc

On a autant de chances de pouvoir sans perte revendre du ferme sur une prime simple pendant le premier cinquième de la durée de l'engagement que pendant les quatre autres cinquièmes.

Pour traiter un exemple particulier, supposons qu'il s'agisse de la rente et que $t = 3$o jours, alors T sera égal à 5 jours. Il y a donc autant de chances, nous apprend la formule, pour que l'on puisse revendre la rente avec l'écart a (28e en moyenne) pendant les cinq premiers jours, que de chances pour qu'on puisse les revendre dans les vingt-cinq jours qui suivent. Parmi les 6o liquidations sur lesquelles portent nos observations, 38 fois l'écart a été atteint : 18 fois pendant les quatre premiers jours, 2 fois pendant le cinquième et 18 fois au delà du cinquième jour.

L'observation est donc d'accord avec la théorie.

Supposons maintenant que $x = 2k\sqrt{t}$, nous trouvons T $= 0,42t$; or la quantité $2k\sqrt{t}$ est l'écart de la prime double, on peut donc dire :

Il y a autant de chances pour que l'on puisse sans perte revendre du

ferme sur une prime double pendant les quatre premiers dixièmes de la durée de l'engagement que pendant les six autres dixièmes.

Occupons-nous encore de la rente : nos observations précédentes nous ont montré que, dans 23 cas sur 60 liquidations, l'écart $2a$ (56e en moyenne) avait été atteint; sur ces 23 cas, l'écart a été atteint 11 fois avant le 14 du mois et 12 fois après cette époque.

L'époque probable serait $0,11\,t$ pour l'option du double et $0,21\,t$ pour l'option du triple.

Enfin, l'époque probable serait la moitié de l'époque totale si x était égal à $2,5\,k\sqrt{t}$.

Distribution de la probabilité. — Nous avons jusqu'à présent résolu deux problèmes :

La recherche de la probabilité à l'époque t.

La recherche de la probabilité pour qu'un cours soit atteint dans un intervalle de temps t.

Nous allons résoudre ce dernier problème d'une façon complète; il ne suffit pas de connaître la probabilité peur que le cours soit atteint avant l'époque t; il faut aussi connaître la loi de probabilité à l'époque t dans le cas où le cours n'est pas atteint.

Je suppose, par exemple, que nous achetions de la rente pour la revendre avec un bénéfice c. Si à l'époque t la revente n'a pu être effectuée, quelle sera, à cette époque, la loi de probabilité de notre opération?

Si le cours c n'a pas été atteint, cela provient de ce que la variation maxima à la hausse a été inférieure à c, alors que la baisse a pu être indéfinie; il y a donc dissymétrie évidente de la courbe des probabilités à l'époque t.

Cherchons quelle sera la forme de cette courbe.

Soit ABCEG la courbe des probabilités à l'époque t, en supposant que l'opération dût subsister jusqu'à cette époque (*fig.* 11).

La probabilité pour que, à l'époque t, le cours c soit dépassé est représentée par l'aire DCEG qui, évidemment, ne fera plus partie de la courbe des probabilités dans le cas de la revente possible.

Nous pouvons même affirmer *a priori* que l'aire de la courbe des probabilités devra encore, dans ce cas, être diminuée d'une quantité égale à DCEG, puisque la probabilité P est le double de la probabilité représentée par DCEG.

Fig. 11.

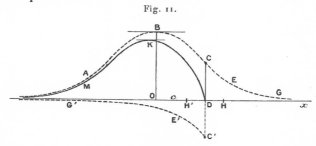

Si le cours c est atteint à l'époque t_1, le cours H aura, à cet instant, la même probabilité que le cours H' symétrique.

La possibilité de la revente au cours c supprime donc, en même temps que la probabilité en H, une probabilité égale en H', et pour avoir la probabilité à l'époque t, nous devons retrancher des ordonnées de la courbe ABC celles de la courbe G'E'C', symétrique de GEC. La courbe de probabilité cherchée sera donc la courbe DKM.

Cette courbe a pour équation

$$p = \frac{1}{2\pi k \sqrt{t}} \left[e^{-\frac{x^2}{4\pi k^2 t}} - e^{-\frac{(2c-x)^2}{4\pi k^2 t}} \right].$$

Cours de probabilité maxima. — Pour obtenir le cours dont la probabilité est la plus grande, dans le cas où le cours c n'a pas été atteint, il suffit de poser $\frac{dp}{dx} = 0$; on obtient ainsi

$$\frac{x}{2c - x} + e^{-\frac{c(c-x)}{\pi k^2 t}} = 0.$$

Si l'on suppose $c = a = k\sqrt{t}$, on obtient

$$x_m = -1,5a,$$

si l'on suppose $c = 2a$, on obtient

$$x_m = -0,4a.$$

Enfin, on obtiendrait

$$x_m = -c,$$

si c était égal à $1,33\,a$.

Cours probable. — Cherchons l'expression de la probabilité dans l'intervalle zéro, u; ce sera

$$\frac{1}{2\pi k\sqrt{t}}\int_0^u e^{-\frac{x^2}{4\pi k^2 t}}\,dx - \frac{1}{2\pi k\sqrt{t}}\int_0^u e^{-\frac{(2c-x)^2}{4\pi k^2 t}}\,dx.$$

Le premier terme a pour valeur

$$\frac{1}{2}\,\Theta\!\left(\frac{u}{2\sqrt{\pi}\,k\sqrt{t}}\right).$$

Dans le second, posons

$$2\sqrt{\pi}\,k\sqrt{t}\,\lambda = 2c - x;$$

ce terme deviendra

$$-\frac{1}{2}\frac{2}{\sqrt{\pi}}\int_0^{\frac{2c}{2\sqrt{\pi}k\sqrt{t}}} e^{-\lambda^2}\,d\lambda + \frac{1}{2}\frac{2}{\sqrt{\pi}}\int_0^{\frac{2c-u}{2\sqrt{\pi}k\sqrt{t}}} e^{-\lambda^2}\,d\lambda.$$

L'expression cherchée de la probabilité est donc

$$\frac{1}{2}\,\Theta\!\left(\frac{u}{2\sqrt{\pi}\,k\sqrt{t}}\right) - \frac{1}{2}\,\Theta\!\left(\frac{2c}{2\sqrt{\pi}\,k\sqrt{t}}\right) + \frac{1}{2}\,\Theta\!\left(\frac{2c-u}{2\sqrt{\pi}\,k\sqrt{t}}\right).$$

Il est intéressant d'étudier le cas où $u = c$ pour connaître la probabilité de bénéfice d'un achat ferme lorsque le cours de revente n'a pu être atteint.

La formule ci-dessus devient dans l'hypothèse $u = c$

$$\Theta\!\left(\frac{c}{2\sqrt{\pi}\,k\sqrt{t}}\right) - \frac{1}{2}\,\Theta\!\left(\frac{2c}{2\sqrt{\pi}\,k\sqrt{t}}\right).$$

Supposons que $c = a$, la probabilité est alors $0,03$.

Si l'écart a n'a jamais été atteint dans l'intervalle t, il n'y a que trois chances sur cent pour que, à l'époque t, le cours se trouve compris entre zéro et a.

On peut acheter une prime simple avec l'idée préconçue de revendre du ferme sur cette prime dès que son écart sera atteint.

La probabilité de la revente est, comme nous l'avons vu, o,69. La probabilité pour que la revente n'ait pas lieu et qu'il y ait bénéfice est o,o3 et la probabilité de perte est o,28.

Supposons que $c = 2a$, la probabilité est alors o,13.

Si l'écart $2a$ n'a jamais été atteint dans l'intervalle t, il y a treize chances sur cent pour que, à l'époque t, le cours se trouve compris entre zéro et $2a$.

Le cours probable est celui dont l'ordonnée divise en deux parties égales l'aire de la courbe des probabilités. Il n'est pas possible d'exprimer sa valeur en termes finis.

Espérance réelle. — L'espérance mathématique $k\sqrt{t} = a$ exprime l'espérance d'une opération qui doit durer jusqu'à l'époque t.

Si l'on se propose de réaliser l'opération dans le cas où un certain écart serait atteint avant l'époque t, l'espérance mathématique a une valeur toute différente, variant évidemment entre zéro et $k\sqrt{t}$ quand l'écart choisi varie entre zéro et l'infini.

Soit c le cours de réalisation d'un achat, par exemple; pour obtenir l'espérance positive réelle de l'opération, on doit ajouter à l'espérance de revente cP l'espérance positive correspondant au cas où la revente n'a pas lieu, c'est-à-dire la quantité

$$\int_0^c \frac{x}{2\pi k\sqrt{t}}\Big[e^{-\frac{x^2}{4\pi k^2 t}} - e^{-\frac{(2c-x)^2}{4\pi k^2 t}}\Big]dx.$$

Si l'on effectue l'intégration du premier terme et si l'on ajoute l'intégrale entière à l'espérance de revente

$$cP = c - c\frac{2}{\sqrt{\pi}}\int_0^{\frac{c}{2\sqrt{\pi}k\sqrt{t}}} e^{-\lambda^2}d\lambda,$$

on obtient pour expression de l'espérance réelle

$$\mathcal{L} = c + k\sqrt{t}\left(1 - e^{-\frac{c^2}{\pi k^2 t}}\right) - c\frac{2}{\sqrt{\pi}}\cdot\int_0^{\frac{c}{\sqrt{\pi}k\sqrt{t}}} e^{-\lambda^2}\,d\lambda,$$

ou

$$\mathcal{L} = c + k\sqrt{t}\left(1 - e^{-\frac{c^2}{\pi k^2 t}}\right) - c\,\Theta\left(\frac{c}{\sqrt{\pi}k\sqrt{t}}\right).$$

Si l'on suppose que $c = \infty$, on retrouve bien $\mathcal{L} = k\sqrt{t}$. On pourrait facilement développer $\mathcal{L}$ en série ; mais la formule qui précède est plus avantageuse, elle se calcule avec les Tables de logarithmes et avec celles de la fonction Θ.

Pour $c = a$, on obtient

$$\mathcal{L} = 0,71\,a;$$

on a de même pour $c = 2a$

$$\mathcal{L} = 0,95\,a.$$

Les espérances de revente étaient, pour ces mêmes écarts, $0,69a$ et $0,86a$.

L'*écart moyen* en baisse, lorsque le cours c n'est pas atteint, a pour valeur

$$\frac{\int_{-\infty}^0 p\,x\,dx}{\int_{-\infty}^0 p\,dx} = \frac{\mathcal{L}}{1 - P - P_1},$$

P_1 désignant la quantité $\int_0^c p\,dx$.

L'écart moyen a donc pour valeur $2,54a$ lorsque $c = a$ et $2,16a$ lorsque $c = 2a$.

Si l'on suppose $c = \infty$, on voit que l'écart moyen est égal à $2a$, résultat déjà obtenu.

Reprenons, à titre d'exemple, le problème général relatif à l'écart a. J'achète ferme avec l'idée préconçue de revendre avec l'écart

$a = k\sqrt{t}$. Si à l'époque t la vente n'a pu être effectuée, je vendrai quel que soit le cours.

Quels sont les principaux résultats que fournit le Calcul des probabilités sur cette opération?

L'espérance réelle positive de l'opération est $0,71\,a$.

La probabilité de la revente est $0,69$.

L'époque la plus probable de la revente est $\dfrac{t}{18}$.

L'époque probable de la revente est $\dfrac{t}{5}$.

Si la revente n'a pas lieu, la probabilité de réussite est $0,03$, la probabilité de perte $0,28$, l'espérance positive $0,02\,a$, l'espérance négative $0,71\,a$; la perte moyenne $2,54\,a$.

La probabilité totale de réussite est $0,72$.

Je ne crois pas nécessaire de présenter d'autres exemples; on voit que la théorie actuelle résout par le Calcul des probabilités la plupart des problèmes auxquels conduit l'étude de la spéculation.

Une dernière remarque ne sera peut-être pas inutile. Si, à l'égard de plusieurs questions traitées dans cette étude, j'ai comparé les résultats de l'observation à ceux de la théorie, ce n'était pas pour vérifier des formules établies par les méthodes mathématiques, mais pour montrer seulement que le marché, à son insu, obéit à une loi qui le domine : la loi de la probabilité.

References

Aleksandrov, P., and A. Kolmogorov. 1936. Endliche Überdeckungen topologis-che Räume. *Fundamental Mathematics* 26:267–271.

Bachelier, L. 1901. Théorie mathématique du jeu. *Annales Scientifiques de l'Ecole Normale Supérieure* 18:143–210.

———. 1912. *Calcul des probabilités*. Paris: Gauthier-Villars.

———. 1939. *Les Nouvelles Méthodes du calcul des probabilités*. Paris: Gauthier-Villars.

———. 1941. Probabilités des oscillations maxima. *Comptes Rendus de L'Acad-émie des Sciences (Paris)* 212:836–838.

Bensoussan, A. 1984. On the theory of option pricing. *Acta Applicandae Math-ematicae* 2:139–158.

Bertrand, J. 1888. *Calcul des probabilités*. Paris: Gauthier-Villars.

Black, F., and M. Scholes. 1973. The pricing of options and corporate liabilities. *Journal of Political Economy* 81:637–654.

Boltzmann, L. 1896. *Vorlesungen über Gastheorie*. Leipzig: J. A. Barth.

Borel, E. 1909. *Éléments de la théorie de probabilités*. Paris: Hermann.

Breiman, L. 1968. *Probability*. Reading, MA: Addison Wesley.

Brown, R. 1828. A brief account of microscopical observations made in the months of June, July and August, 1827, on the particles contained in the pollen of plants; and on the general existence of active molecules in organic and inorganic bodies. *Philosophical Magazine* 4(21):161–173.

Bru, B., and M. Yor. 2002. Comments on the life and legacy of Wolfgang Doeblin. *Finance and Stochastics* 6:3–47.

Bucy, R., and P. Joseph. 1968. *Filtering for Stochastic Processes, with Applications to Guidance*. Wiley Interscience.

Bucy, R., and R. Kalman. 1961. New results in linear filtering and prediction theory. *Journal of Basic Engineering ASME* 81:95–108.

Ciesielski, Z. 1961. Hölder conditions for realizations of Gaussian processes. *Transactions of the American Mathematical Society* 99:403–413.

Conway, F., and J. Siegelman. 2005. *Dark Hero of the Information Age*. New York: Basic Books.

Cootner, P. 2001. *The Random Character of Stock Market Prices*. London: Risk Books. (Reprinted from the original 1964 edition published by MIT Press.)

Courrège, P. 1963. Intégrales stochastiques et martingales de carré intégrable. *Séminaire Brélot-Choquet-Dény (Théorie du Potentiel)*, 7ème année 1962-63.

Courtault, J.-M., and Y. Kabanov. 2002. *Louis Bachelier: aux origines de la finance mathématique.* Besançon: Presses Universitaires Franc-Comtoises.

Courtault, J.-M., Y. Kabanov, B. Bru, P. Crépel, I. Lebon, and A. Le Marchand. 2000. Louis Bachelier on the centenary of *Théorie de la spéculation. Mathematical Finance* 10:341–353.

Cox, J., and S. Ross. 1976. The valuation of options for alternative stochastic processes. *Journal of Financial Economics* 3:145–166.

Cox, J., S. Ross, and M. Rubinstein. 1979. Option pricing: a simplified approach. *Journal of Financial Economics* 7:229–263.

Dale, R. 2004. *The First Crash.* Princeton University Press.

de Moivre, A. 1738. *The Doctrine of Chances, or, a Method of Calculating the Probabilities of Events in Play,* 2nd edn. Printed for the author by H. Woodfall, London.

de Montessus, R. 1908. *Leçons élémentaires sur le calcul des probabilités.* Paris: Gauthier-Villars.

de Seingalt (Giacomo Girolamo), C. 1922. *Memoirs* (trans. A. Machen), Volume 4. London: Casanova Society.

Delbaen, F., and W. Schachermayer. 1994. A general version of the fundamental theorem of asset pricing. *Mathematische Annalen* 300:463–520.

Döblin, A. 1929. *Berlin Alexanderplatz.* Olten: Walter.

Doeblin, W. 1938. Sur l'équation de Kolmogoroff. *Comptes Rendus de L'Académie des Sciences (Paris)* 207:705–707.

——. 2000. Sur l'équation de Kolmogoroff. Pli cacheté déposé le 26 février 1940, ouvert le 18 mai 2000. *Comptes Rendus de L'Académie des Sciences (Paris), Série I* 331:1031–1187.

Doeblin, W., and P. Lévy. 1936. Sur les sommes de variables aléatoires indépendantes à dispersions bornées inférieurement. *Comptes Rendus de L'Académie des Sciences (Paris)* 202:2027–2029.

Doléans-Dade, C., and P.-A. Meyer. 1970. Intégrales stochastiques par rapport aux martingales locales. *Séminaire de Probabilités,* IV. Lecture Notes in Mathematics, Volume 124, pp. 77–107.

Doob, J. L. 1937. Stochastic processes depending on a continuous parameter. *Transactions of the American Mathematical Society* 42:107–140.

——. 1942. The Brownian movement and stochastic equations. *Annals of Mathematics* 43:351–369.

——. 1945. Markoff chains: denumerable case. *Transactions of the American Mathematical Society* 58:455–473.

——. 1953. *Stochastic Processes.* John Wiley.

——. 1970. William Feller and twentieth century probability. In *Sixth Berkeley Symposium on Mathematical Statistics and Probability,* Volume 2, pp. xv–xx.

——. 1972. Obituary: Paul Lévy. *Journal of Applied Probability* 9:870–872.

Dynkin, E. B. 1989. Kolmogorov and the theory of Markov processes. *Annals of Probability* 17:822–832.

Einstein, A. 1905. Über die von der molekularkinetischen Theorie der Wärme geforderte Bewegung von in ruhenden Flüssigkeiten suspendierten Teilchen.

(On the motion of particles suspended in fluids at rest implied by the molecular-kinetic theory of heat.) *Annalen der Physik* 17:132–148.

Einstein, A. 1906. Zur theorie der Brownschen Bewegung. (On the theory of Brownian motion.) *Annalen der Physik* 19:371–381.

Feller, W. 1936. Zur Theorie der Stochastischen Prozessen (Existenz und Eindeutigkeitssätze). *Mathematische Annalen* 113:113–160.

——. 1950. *An Introduction to Probability Theory and Its Applications*, Volume 1. John Wiley.

——. 1951. Diffusion processes in genetics. In *Proceedings of the Second Berkeley Symposium on Mathematical Statistics and Probability*, pp. 227–246.

——. 1966. *An Introduction to Probability Theory and Its Applications*, Volume 2. John Wiley.

Ferguson, N. 2001. *The Cash Nexus*. Penguin.

Föllmer, H., and A. Schied. 2004. *Stochastic Finance*. Berlin: Walter de Gruyter.

Girsanov, I. 1960. On transforming a certain class of stochastic processes by absolutely continuous substitution of measures. *Theory of Probability and Applications* 5:285–301.

Hadamard, J. 1943. Obituary: Émile Picard. *Journal of the London Mathematical Society* 18:114–128.

Harrison, J., and D. Kreps. 1979. Martingales and arbitrage in multiperiod securities markets. *Journal of Economic Theory* 20:381–408.

Harrison, J., and S. Pliska. 1981. Martingales and stochastic integrals in the theory of continuous trading. *Stochastic Processes and Their Applications* 11:215–260.

Homer, S., and R. Sylla. 2005. *A History of Interest Rates*, 4th edn. Wiley.

Hull, J. 2005. *Options, Futures and Other Derivatives*, 6th edn. Prentice Hall.

Hunt, G. 1956. Some theorems concerning Brownian motion. *Transactions of the American Mathematical Society* 81:249–319.

Itô, K. 1944. Stochastic integral. *Proceedings of the Imperial Academy, Tokyo* 20:519–524.

——. 1951. *On Stochastic Differential Equations*. Memoirs of the American Mathematical Society, Volume 4. Providence, RI: American Mathematical Society.

——. 1986. *Kiyosi Itô Selected Papers* (ed. D. W. Stroock and S. R. S. Varadhan). Springer.

Ito, K., and H. McKean, Jr. 1965. *Diffusion Processes and Their Sample Paths*. Springer.

Jacod, J., and M. Yor. 1977. Étude des solutions extrémales et représentation intégrale des solutions pour certaines problèmes de martingales. *Zeitschrift für Wahrscheinlichkeitstheorie und verwandte Gebiete* 38:83–125.

Jamshidian, F. 1997. LIBOR and swap market models and measures. *Finance and Stochastics* 1:293–330.

Jarrow, R., and P. Protter. 2004. *A Short History of Stochastic Integration and Mathematical Finance: The Early Years, 1880–1970*. IMS Lecture Notes Monograph, Volume 45, pp. 1–17.

185

Kakutani, S. 1944a. Markoff processes and the Dirichlet problem. *Proceedings of the Japanese Academy* 21:227–233.

——. 1944b. Two-dimensional Brownian motion and harmonic functions. *Proceedings of the Imperial Academy, Tokyo* 20:706–714.

Kalman, R. 1960. A new approach to linear filtering and prediction problems. *Journal of Basic Engineering, ASME* 82:35–45.

Karatzas, I. 1988. On the pricing of American options. *Applied Mathematics and Optimization* 17:37–60.

Karatzas, I., and S. Shreve. 1991. *Brownian Motion and Stochastic Calculus.* Springer.

——. 2001. *Methods of Mathematical Finance.* Springer.

Kolmogorov, A. 1931. Über die analytischen Methoden in der Wahrscheinlichkeitsrechnung. *Mathematische Annalen* 104:415–458.

——. 1933. *Grundbegriffe der Wahrscheinlichkeitsrechnung.* Springer.

Kunita, H., and S. Watanabe. 1967. On square-integrable martingales. *Nagoya Mathematical Journal* 30:209–245.

Kushner, H. 1967. *Stochastic Stability and Control.* Academic Press.

Langevin, P. 1908. Sur la théorie du mouvement brownien. *Comptes Rendus de L'Académie des Sciences (Paris)* 146:530–533.

Laplace, P. S. 1810. Mémoire sur les approximations des formules qui sont fonctions de très grands nombres, et sur leur application aux probabilitiés. *Mémoires de l'Académie des Sciences, 1ère Série* X:301–345.

——. 1812. *Théorie Analytique des Probabilités,* 1st edn (2nd edn, 1814; 3rd edn, 1820). Paris: Courcier.

Lauritzen, S. 2002. *Thiele: Pioneer in Statistics.* Oxford University Press.

Lefèvre, H. 1853. *Traité des valeurs mobilières et des operations de bourse: placement et spéculation.* Paris: E. Lachaud.

Lévy, P. 1937. *Théorie de l'addition des variables aléatoires.* Paris: Gauthier-Villars.

——. 1948. *Processus stochastiques et mouvement brownien.* Paris: Gauthier-Villars.

——. 1955. Wolfgang Doeblin (V. Doblin) (1915–1940). *Revues d'histoire des sciences et leur applications* 8:107–115.

Lipton, A. 2003. *Exotic Options: The Cutting Edge Collection.* London: Risk Books.

Loève, M. 1973. Paul Lévy: 1886–1971. *Annals of Probability* 1:1–8.

McKean, H. P. 1969. *Stochastic Integrals.* Academic Press.

Markov, A. 1906. Extension of the law of large numbers to dependent events. *Izvestiya Fiziko-matematicheskogo obschestva pri Kazanskom universitete, 2-ya seriya* 15:135–156.

——. 1908. *Extension of Limit Theorems of Probability to a Sum of Variables Connected in a Chain.* Notes of the Imperial Academy of Sciences of St Petersburg VIII, Series Physio-Mathematical College, Volume XXII, No. 9. Imperial Academy of Sciences of St Petersburg.

Markowitz, H. 1952. Portfolio selection. *Journal of Finance* 7:77–91.

Merton, R. 1973. Theory of rational option pricing. *Bell Journal of Economics and Management Science* 4:141–183. (Chapter 8 of Merton (1992).)

———. 1992. *Continuous-Time Finance.* Basil Blackwell.

Meyer, P.-A. 1962. A decomposition theorem for supermartingales. *Illinois Journal of Mathematics* 6:193–205.

———. 1963. Decomposition of supermartingales: the uniqueness theorem. *Illinois Journal of Mathematics* 7:1–17.

———. 1966. *Probability and Potentials.* Waltham, MA: Blaisdel.

Musiela, M., and M. Rutkowski. 2004. *Martingale Methods in Financial Modelling,* 2nd edn. Springer.

Paley, R., and N. Wiener. 1934. *Fourier Transforms in the Complex Domain.* American Mathematical Society Colloquium Publications, Volume XIX. Providence, RI: American Mathematical Society.

Paley, R., N. Wiener, and A. Zygmund. 1933. Notes on random functions. *Mathematische Zeitschrift* 37:647–668.

Perrin, J. 1909. Mouvement brownien et réalité moléculaire. *Annales de Chimie et de Physique* 18:1–114.

———. 1912. *Les Atomes.* Paris: Felix Alcan.

Preda, A. 2004. Informative prices, rational investors: the random walk hypothesis and the nineteenth-century 'science of financial investments'. *History of Political Economy* 36:351–386.

Press, W., S. Teukolsky, W. Vetterling, and B. Flannery. 2001. *Numerical Recipes in C++.* Cambridge University Press.

Ray, D. 1956. Stationary Markov processes with continuous paths. *Transactions of the American Mathematical Society* 82:452–493.

Rayleigh, J. 1880. On the resultant of a large number of vibrations of the same pitch and arbitrary phases. *Philosophical Magazine* 10(5):73–78.

Regnault, J. 1853. *Calcul des chances et philosophie de la bourse.* Paris: Mallet-Bachelier et Castel.

Russell, B. 1967. *The Autobiography of Bertrand Russell,* Volume 1, pp. 1872–1914. Sydney: Allen and Unwin.

Samuelson, P. 1965. Rational theory of warrant pricing. *Industrial Management Review* 6:13–39.

Shiryaev, A. N. 1966. On stochastic equations in the theory of conditional Markov processes. *Theory of Probability and Its Applications* 11:179–184.

Smoluchowski, M. 1906. Zur kinetischen Theorie der Brownschen Molekularbewegung und der Suspensionen. *Annalen der Physik* 21:756–780.

Snell, L. 2005. Obituary: Joseph Leonard Doob. *Journal of Applied Probability* 42:247–256.

Taqqu, M. S. 2001. Bachelier and his times: a conversation with Bernard Bru. *Finance and Stochastics* 5:3–32.

Thackeray, W. M. 1848. *Vanity Fair.* London: Bradbury & Evans.

Ville, J. 1939. *Étude critique de la notion de collectif.* Paris: Gauthier-Villars.

Wiener, N. 1923. Differential space. *Journal of Mathematical Physics* 2:127–146.

Wiener, N. 1949. *Extrapolation, Interpolation and Smoothing of Stationary Time Series.* John Wiley.

———. 1953. *Ex-Prodigy: My Childhood and Youth.* Cambridge, MA: MIT Press.

Wong, E., and M. Zakai. 1965. On the relation between ordinary and stochastic differential equations. *International Journal of Engineering Science* 3:213–229.